Fourth Edition

Experience the Extraordinary Chemistry of Ordinary Things

A LABORATORY MANUAL

B. Coburn Richardson

Highline Community College
Des Moines, Washington

Thomas G. Chasteen

Sam Houston State University
Huntsville, Texas

WILEY

Cover Photo: Steve Lupton/CORBIS.

To order books or for customer service call 1-800-CALL-WILEY (225-5945).

ISBN 978-0471-42358-4

Printed in the United States of America

SKY10029213_081921

Printed and bound by Sheridan, KY, A CJK Group Company

Acknowledgments

Teacher-authors are always in great debt to the two "S's"—students and shoulders. For thirty years, students in "Chemistry for Those Who Hate Chemistry" classes at Highline Community College have unwittingly served as guinea pigs for many of the experiments in this manual. A much smaller group has helped test experiments at Sam Houston State University. Our students continue to teach us how to design and write experimental procedures for the non-scientist. It has been a long and humbling but rewarding challenge.

While at least a half dozen experiments in this manual have never appeared before insofar as these authors are aware, no writer ever works in an intellectual vacuum. Many ideas have come from standing on the shoulders of our contemporaries and of those who have gone before. We would especially like to thank Rubin Battino, James P. Birk, Arlo D. Harris, Martin G. Ondrus, and Michael Sady (and their coauthors) for their excellent *Journal of Chemical Education* articles from which five of these labs were drawn. Thank goodness that chemistry is a cumulative science and that those that have gone before (even just before) leave recorded tracks.

We would also like to thank the many reviewers for the first and subsequent editions of this work. Many of the suggested additions and changes have been included. Now this fourth edition has been undertaken with two goals in mind:

> 1. to clarify procedural items that might still cause some students possible confusion. We strive to do this every time we revise our work; our goal remains for non-science students to be able to follow the written and pictorial directions with a very minimum of outside help. To this end, 46 new pictures, drawings and tables alone have been added to this new edition.

> 2. to update introductions and background sections to reflect recent changes.

We strongly feel that these revisions and additions to this fourth edition will help students more clearly understand procedures and experimental apparatus; more easily get results; and increase student success in the laboratory.

This manual is indeed a living organism evolving in response to its surroundings—the students that made it all happen. We thus solicit and welcome the continued suggestions from all who use this book and hereby promise a reply to all of you. And we thank our wives, Sharon and Tamara, for their understanding and support so that we can devote all energies necessary to make this text the best that we are able. And thanks to those who have contacted us with questions and suggestions from "out in the field."

Tom Chasteen
Sam Houston State University
CHM_TGC@SHSU.EDU
http://www.shsu.edu/~chm_tgc

Bruce Richardson
Highline Community College
BRICHARD@HIGHLINE.EDU
http://flightline.highline.edu/brichardson/

Authors Biographies

Dr. Bruce Coburn Richardson received his formal education at San Diego State U., Oregon State U., Fana Folkehøgskule (Norway) and Univ. of Colorado. After a few years of industrial research, he joined Highline College near Seattle "at half his former salary but twice the fun"! Bruce has especially enjoyed the challenges of developing manuals for a liberal arts course *"Chemistry for Those Who Hate Chemistry"* which he has taught since 1972, as well as *Organic Chemistry Laboratory* and *Spectrometric Identifications*.

Other interests include hiking, folk music and dance, table tennis, photography, southern gospel music, backhoes, sailing, computers, playing the accordion and classic guitar. He shepherds a small sheep farm with his wife and light of 34 years, Sharon Noel.

"We live in such flagrant luxury by world standards and take *so* much *too* much for granted, both materially and those we love. *Yesterday is history, tomorrow is a mystery and today is the present — because it is indeed a gift.* Cherish it."

Thomas Chasteen completed a B.S. and M.S. at East Texas State University. After a 3 1/2 year stint in industry involving positions involving bench chemistry, lab coordinator, and ultimately director of research and development at a small company, he went on to a Ph.D. at the University of Colorado at Boulder under John Birks. He completed his doctorate in 1990 with work involving the determination of selenium and tellurium in biological headspace using fluorine-induced chemiluminescence.

He lives in Texas with his wife Tamara and sons Charles and Peter. Now professor of chemistry at Sam Houston State University he teaches analytical and instrumental chemistry, environmental studies, and honors courses for nonscience majors. Among others, his research interests include the determination of selenium and tellurium bioprocessed by bacteria.

Preface

This preface explains why your chemistry course has a lab, and why the lab is structured in its particular way. In answer to the first question, we must hasten to emphasize that, unlike liberal arts subjects like philosophy, chemistry is strictly an *experimental* science. This is not to imply that chemical knowledge has always progressed and expanded in an orderly fashion based upon scientific observations and experiments. Much chemical "thought" 2000 or so years ago evolved from an "armchair philosophy" rather than from deduction based upon careful experimentation. The result of this was that untruths and misconceptions retarded progress towards a real understanding of the matter/energy make-up of our world and universe for many centuries. The most noted fallout from this was perhaps the impossible attempts of alchemists to transmute (change) lead into gold by chemical reactions.

It is thus primarily through experimentation—especially quantitative experimentation where not only what happens is noted, but also how much of what—that the laws and concepts of chemistry have been clarified. Since such experimentation forms the basis of all our chemical knowledge, no chemistry course can really convey what chemistry is all about unless we have the chance to try our own handiworks with some "live" molecules, however modest the effort.

The lab is where the action is, and it is hoped that it can and will make chemistry come alive for you and perhaps be that one aspect of the course that may linger longest in your memory (favorably, we hope). Some of you will be able to get good results in lab and some will have difficulty, but whatever your success or lack of it, you will know more about that particular aspect of a chemist's life. The nature of your results will not affect your lab grade. Chemistry can, but need not be, traumatic, numbing and nerve-wracking, and it can also be fun. This is one of the reasons why there are so many photographs: they are meant to help you determine just how things look, how glassware or the apparatus fits together and even "how some reactions go" before you experiment yourself.

Now what about the design of the lab? In spite of the fact that this will be the first (and perhaps the last) chemistry course you'll ever take, the lab procedures in large measure do not follow a set "cookbook" of steps that everyone follows to the same supposed inevitable end. The techniques themselves will be described in considerable detail and are illustrated with photographs and drawings of apparatus and equipment, but each person or group may well be working on a different sample. This approach will demand extra time and responsibility on both your and your lab instructor's part, but these authors are willing to wager that you will come to agree the effort is worth it.

The basic objectives which guided the form and selection of experiments are the following:

1. Relate chemistry to that part of the world that is meaningful and/or familiar.

2. Stress sample-from-home type experiments.

3. Guarantee reasonable chance of experimental success.

4. Provide challenging experiments.

5. Maximize quantitative types of investigations.

6. Make any computations required relatively simple to perform without knowledge of algebra.

7. Cover as wide a variety of topics and techniques as possible.

8. Use experiments that can be performed (at least the laboratory part) within two hours.

9. Avoid toxic chemicals or hazardous procedures wherever possible.

10. Permit checking accuracy of results wherever possible by providing known samples or other necessary information.

11. Give very detailed experimental procedures, allowing for a variety of student-chosen samples, which take into account probable lack of scientific backgrounds and experience.

12. Give historical backgrounds and explain rationale for procedural steps—give the "whys" as well as the "whats."

The experiments selected are covered in depth and some use procedures rather sophisticated at the beginning level—but nonetheless adaptable by most students. A detailed Instructor's Manual is available from John Wiley containing necessary information about reagent preparation, supply requirements, trouble shooting tips, etc. for the lab's instructors/teaching assistants.

We mentioned responsibilities; you must agree and be willing to fulfill two duties if we are to succeed. First, you must attend lab faithfully; or to put it more candidly, lab attendance is required.

Second, you will find that many of the experiments require you to bring your own sample from home. Sure, we could provide you and everyone else with identical samples to test with results almost guaranteed. But that is not the intent; you are given the opportunity to pick, within limits of the type of experiment, a sample of your choice. It is your decision, your opportunity, but also your responsibility to bring the sample that is specified at the beginning of each lab—so don't blow it. A "forgotten" sample necessary for an analysis may mean you cannot do the experiment.

We now welcome you to our chemical lair—and the chance to enjoy some "Close encounters of the Chemical Kind." Don't forget your goggles....

Bruce Richardson   Red Chasteen
November 2002

Contents

Experience
The Extraordinary
Chemistry
of
Ordinary Things

A Laboratory Manual
Fourth Edition

Introduction

First Day in the Laboratory

Check-in

Check your drawer equipment against the supplied check sheet with the help of pictures at the end of this "experiment." Note any missing, broken or dirty equipment and get replacements from the stockroom.

Discussion

We don't want to put the fear of chemistry in you (anymore than it may already be!) by dwelling on the unpleasant, but as a metric equivalent of an English saying might go, "Thirty grams of prevention is worth 0.454 kilograms of cure." Actually, working in the chemistry lab may be safer than working at home. For example, many of you probably have a chemical in your kitchen potentially more hazardous than anything you can see around you here in the lab—the aerosol can of oven cleaner containing sodium or potassium hydroxide (see Experiment #14: *Label Reading*)! So now here we come with some chemistry "rules."

A. Surviving in the Lab by Not Blowing Yourself Up
(Laboratory Safety and Techniques Check List)

Safety

1. Our eyes are very easily damaged and thus are of utmost concern in the laboratory. Eye protection will be mandatory for most experiments. This means that you must wear approved safety glasses or goggles whenever you are in the laboratory area, irrespective of what you may be doing at the time. (Realize that other student's experiments may *erupt* on you at any time.)

If something should splash into your eyes, move FAST—seconds really count. Make sure you know in advance the location and use of the eye wash fountain as well as the safety shower. Flush both of your eyes with the eye wash water until your lab instructor tells you to stop.

2. If you spill a chemical on your skin or clothing, wash it immediately with plenty of water and consult your lab instructor. If the spilled liquid is an acid or base (the common situation), neutralize the spill first with the appropriate neutralizing solution and then follow with a water wash.

You may slosh these neutralizing solutions liberally over the affected area. These solutions are very safe liquids, but do not put them into the eyes: sodium bicarbonate (baking soda solution) for zapping acids; acetic acid (strong vinegar) for zapping bases. How do we know we have spilled an acid? (*Answer:* look for the word *acid* on the bottle. Easy, huh.) But how do we know we have spilled a base? Gotcha! The word *base* is not on the bottle. (*Answer:* look for the word *hydroxide* on the bottle.) Everything can't always be simple in chemistry!

3. If diluting sulfuric acid, pour the acid slowly into the water, stirring constantly. Never add water to the acid, since so much heat is liberated that steam may be formed with almost explosive violence.

4. NEVER taste a chemical or solution; poisonous substances are not always so labeled in the laboratory. It is always best to follow the credo "Guilty unless proven innocent" when considering the potential hazard from a chemical.

5. When smelling the odor of a liquid, do not hold your face directly over the container, lest you chance an unpleasant surprise. (Would you jam your nose into a bottle of household cleaning ammonia?) Instead, fan a little of the vapor towards you (called *wafting*) by sweeping your hand over the top of the container towards your nose.

6. To protect your clothing from corrosive chemicals, an inexpensive plastic apron (much cheaper than new clothing) is strongly recommended.

7. Coats, bags, rucksacks etc. should not be piled onto the bench tops nor dumped onto the floor. Not only might they pick up corrosive chemicals this way, but they can be a safety hazard. Use the coat racks or storage bins.

A typical shower/eyewash station.

8. When putting glass tubing through a rubber stopper, first put a drop of water or glycerine on the tubing and on the hole in the stopper. Hold the tubing with a cloth near the end to be inserted and insert with a twisting motion. *Don't force*. Accidents of this kind are the most common way hands become cut or impaled.

9. Do not point your test tube at your neighbor or yourself when heating substances. Sudden boiling may turn your test tube into a scalding liquid cannon!

10. Note the location of fire extinguishers and fire blankets, as well as the safety shower and eye wash fountains. Know how to use them.

Techniques

1. If you are used to throwing all of your household wastes into the same container (shame on you!), then you will have to pay particular attention in the laboratory. The sorting of wastes is not only responsible management, but potential fires between incompatible chemicals make this imperative.

Put all solid chemicals to be discarded into designated waste crocks in the laboratory and replace the top securely. Put ordinary waste paper into suitably marked receptacles. Empty liquids into appropriately labeled containers unless told by your lab instructor that they may be flushed down the sink drain with running water. Never throw matches, filter paper or any solid waste into the sink or troughs.

2. Bottles containing liquids generally have a stopper designed to be removed in a specific fashion. The best technique avoids laying this stopper down at all. Instead, pull the stopper out by placing the *back* of your hand over the stopper and grasping it between the middle and forefinger. This same hand can then be used to pick up the bottle as well, leaving your other hand free to hold the container into which you are pouring the liquid. Ignoring this technique could lead to contamination of the bottle contents.

To escape the wrath of your lab instructor, if you must put a stopper (from a bottle of liquid) down on the bench top, be sure that its tip does *not* touch the surface. If by accident it does touch, wash it off using a stream of distilled water from a wash bottle. Many students use these chemicals, and we must take every precaution to keep them from becoming contaminated. The bench top area should also be wiped clean to remove any of the liquid contents which might have been deposited from the stopper onto the surface.

3. Do not insert your own pipettes, medicine droppers or spatulæ into the reagent bottles. Instead, pour out a little of the chemical into a small container and then take your sample from this container. This is done also to avoid possible contamination.

4. NEVER return unused chemicals to the stock bottles. This might sound very wasteful, but someone may make a mistake from which many other student experiments could suffer. If you take too much of a chemical, either share it with others or discard what is left into an appropriately labelled container.

5. Weigh solids on a paper, plastic boat or a watch glass. Do not allow chemicals to come into contact with the balance pans to avoid contamination and corrosion.

6. Keep your working area orderly and always leave it clean for the next person at all times.

7. Be forewarned that, however clean looking, bench tops may have chemical residues lurking on them ready to eat into anything they come into contact with: clothes, book bags, etc.

B. Surviving in the Lab by Making Your Instructor Happy
(Rules of the Game)

1. You will receive specific directions for keeping your lab instructor happy, which may include:

> (a) handling of lab makeups;
> (b) due dates for turning in report/question sheets;
> (c) handling of late turn-ins;
> (d) report sheet points;
> (e) how lab points affect course grade; and
> (f) group work on experiments.

2. Lab attendance is required.

3. A helpful (but not plagiaristic) collusion among yourselves is encouraged in completing the report sheets and answering questions at the end of the experiment.

4. When samples from home are analyzed, none of your score will be based upon how "good" your results are.

5. Answers to questions at the end of each experiment can be found in:

> (a) the background discussion section to the experiment,
> (b) a chemistry text,
> (c) a dictionary or
> (d) your creative thinking.

Safety Symbols

The experimental procedures that appear in this manual will be accompanied by safety symbols integrated into the text. These symbols are included to alert you, the experimenter, to that part of the procedure that requires extra care and attention. The specific safety symbols refer to specific dangers that may be encountered. For instance, the corrosive safety symbol will appear near the point in the procedure in which you are directed to use a strong acid or base or other corrosive material that requires care in handling.

The safety symbols that are used in this manual are listed on the next page.

Corrosive Material

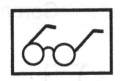

Eye Protection Required

Fire Hazard

Poisonous Material

Finally, this instructor icon will alert you to a step where you may need help from the lab instructor or may need to show that person your equipment set-up.

Common Laboratory Equipment

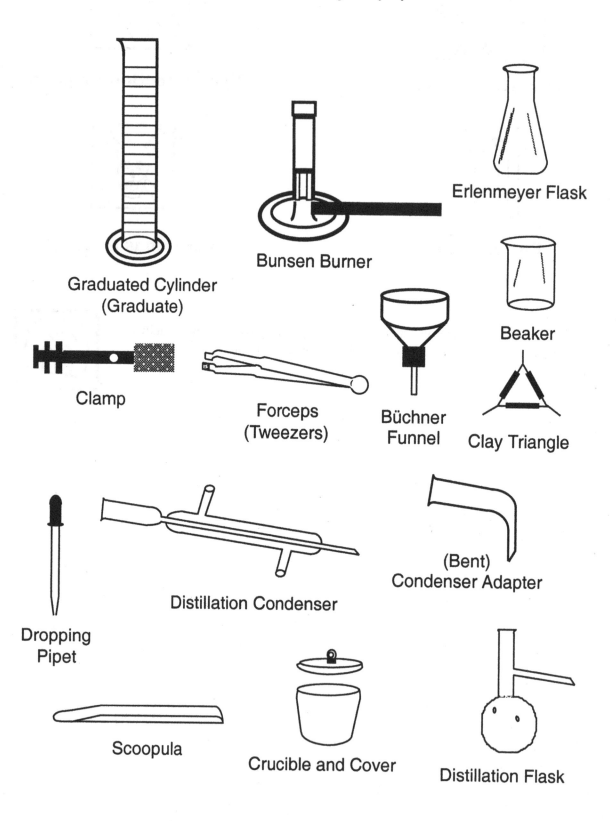

Graduated Cylinder
(Graduate)

Bunsen Burner

Erlenmeyer Flask

Clamp

Forceps
(Tweezers)

Büchner
Funnel

Beaker

Clay Triangle

Dropping
Pipet

Distillation Condenser

(Bent)
Condenser Adapter

Scoopula

Crucible and Cover

Distillation Flask

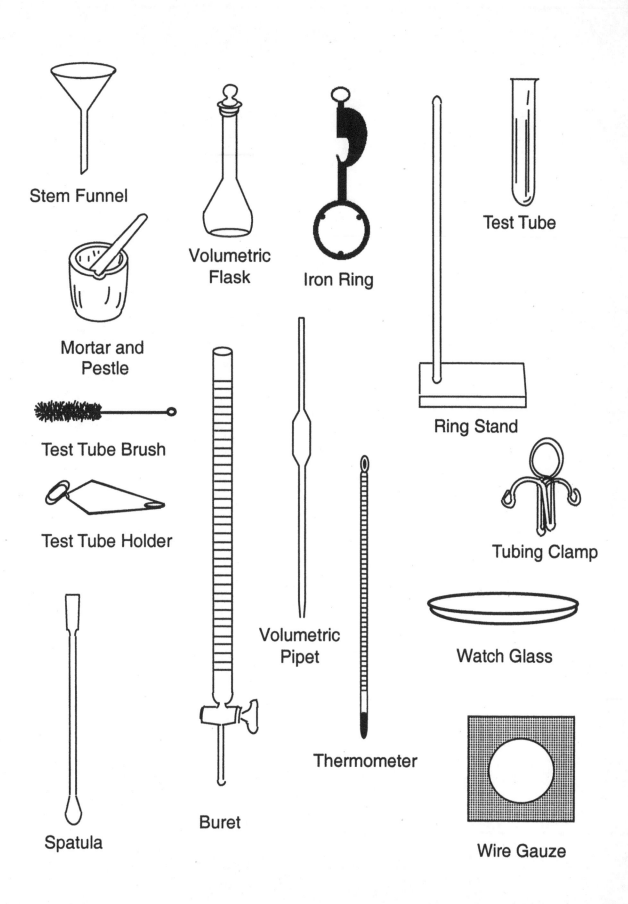

Stem Funnel

Mortar and
Pestle

Test Tube Brush

Test Tube Holder

Spatula

Volumetric
Flask

Buret

Iron Ring

Volumetric
Pipet

Thermometer

Test Tube

Ring Stand

Tubing Clamp

Watch Glass

Wire Gauze

Experiment 1

The Ubiquitous Bunsen Burner

Sample From Home

No samples from home are needed for this experiment.

Objectives

You will examine the structure and function of the Bunsen burner and the effects that the air and gas flow controls have upon the color and hotness of the flame. Time and materials will also be available for experimenting with glassworking.

Background

Those pieces of laboratory equipment that visions of the proverbial white-smocked, bespectacled chemist conjures up in most people's mind would have to be the test tube—and the Bunsen burner. What self-respecting chemist indeed would be caught in any Hollywood movie without them! Although the Bunsen burner has been around ever since Robert Bunsen developed it back in 1855, it is still a workhorse in the lab for many heating jobs that require neither careful heat control, nor involve flammable chemicals.

Burning is a chemical process and involves a chemical reaction whereby a substance commonly combines with oxygen in the air to produce new substances. These new substances, while containing all of the same original atoms, have entirely new physical and chemical properties. None of the atoms representing elements have changed their fundamental identity, but they have undergone a chemical change which is accompanied by drastic alterations in their properties. Thus the paraffin wax in a candle "burns" (reacts with oxygen) to produce carbon dioxide gas, water, light and heat energy. Decaying organic matter such as leaves and humus also "burn," only not usually so fast that the heat is sufficient to cause ignition. Iron, too, slowly "burns" in air to produce rust—a new chemical compound of iron combined with oxygen.

The actual substance burned in the Bunsen burner is principally a gaseous paraffin hydrocarbon called methane (marsh gas or natural gas). A chemically correct equation for this reaction would read: methane and oxygen in air react to produce carbon dioxide, water and energy. In chemist's formula notation, this would become:

$$CH_4 \;+\; 2O_2 \;\longrightarrow\; CO_2 \;+\; 2H_2O \;+\; \Delta \;+\; h\sqrt{}$$

| 1 molecule methane | 2 molecules oxygen | 1 molecule carbon dioxide | 2 molecules water | heat | light |

The actual design of the Bunsen burner is in principle the same as for other types of gas flames, whether in a home gas stove or a welding torch. The purpose of this design is to premix the gas and oxygen (air) before ignition. This permits the combustion process to be more efficient, thereby producing more heat as well as a particular "hot zone" defined by the shape of the burner mouth. The flame color is an indication of how complete the burning has been. A yellow color in the flame is due to microscopic incandescent carbon particles (soot); this condition is undesirable because any carbon atoms escaping as soot and not ending up as carbon dioxide reduce the efficiency and hence heat of the flame. But these are things you are to observe and verify in this experiment.

Procedure

1. Determine where the gas control is for your particular type of burner.
Some may have a thumbscrew control underneath the base, but in any case the gas flow can always be adjusted back at the main valve at your desk station. The air control consists of vents at the base of the tall cylindrical chimney and can be opened and closed either with a circular slip-ring or by rotating the chimney itself. Turn on the gas and light the Bunsen burner. If the flame keeps blowing itself out, cut back a bit on the gas flow valve.

2. *Shut off the air vents.* Screw the vertical burner tube ("chimney") firmly down onto its base. If your burner instead has a slip ring at the bottom of the chimney, the air flow can be shut off completely only by tightly gripping the bottom of the chimney and wrapping your hand around it using your ring and little fingers. Yes, it is safe to handle the Bunsen burner while it is lit if you are careful. But you MUST shut off the air vents to obtain valid observations. **Repeat: you must shut off the air vents to obtain valid observations**. Any

blueness to your flame means that you have *not* completely shut off the air. Your lab instructor can demonstrate.

(a) You should now have seen a dramatic change in the nature of the flame. Note the color and shape of your flame on the report sheet.

(b) Briefly (20–30 seconds) using your tongs, hold a white porcelain evaporating dish filled with cold water over the very top of this flame and observe any deposit formed on the underside of the dish.

12

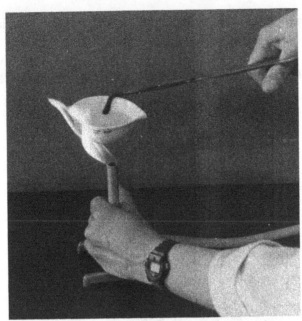

Figure 1.1. Examining no-air flame using an evaporating dish.

(c) Hold your wire gauze (not a thermometer) for about 10 seconds at various levels in the flame in a horizontal position and pass it up and down through the flame. Does the wire glow red hot at any location in the flame? (If the gauze has a white ceramic center, heat the bare metal at the corners.)

3. *Now completely open the air vents.* If your flame goes out, you may have to cut back slightly on the gas flow and relight the burner.

(a) You should at this point be able to discern a relatively colorless flame with two distinct conical sections. If in doubt, ask for help from your lab instructor.

(b) Repeat the heating test with your evaporating dish as in 2(b).

(c) Check the hotness (temperature) of your flame as in 2(c).

(d) Draw a temperature profile sketch, showing the general form of the flame and its coldest and hottest parts.

4. (Optional). Your lab instructor can demonstrate how to seal the end of glass tubing and then blow out (not *blow up!*) sections of this glass.

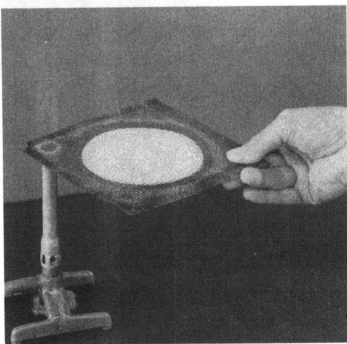

Figure 1.2. Examining temperature of air-mix flame using a a wire gauze.

Help yourself to additional pieces of glass tubing and try your hand at some glassworking. Feel free to take home whatever you blow, bend, or stick together.

REMEMBER: EYE PROTECTION IS MANDATORY WHEN WORKING IN THE LABORATORY; THIS NOW INCLUDES DOING ANY GLASSBLOWING.

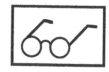

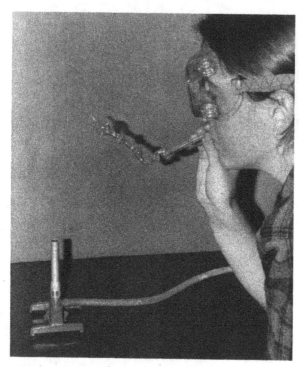

Figure 1.3. Creative glassblower at work.

The Bunsen Burner

(Will you remember this old flame from your chemistry daze?)

Date _____   **Section number** _____   **Name** _____

1. Observations on Bunsen burner set-up

    Gas control located _____.

    Air control located _____.

2. Air vents closed

    (a) Appearance of flame _____.

    (b) Nature of deposit, if any, on dish (identify) _____.

    (c) Did your gauze glow red anywhere in flame? _____.

3. Air vents open

    (a) Appearance of flame _____.

    (b) Nature of deposit, if any, on dish _____.

    (c) Did your gauze glow red anywhere in flame? _____.

    (d) Draw the shape and temperature profile sketch of your flame below.

The Bunsen Burner

Date _____   **Section number** _____ **Name**_____

1. Does soot represent an element or a compound? Explain.

2. Where do the carbon atoms come from in this experiment that are found in soot?

3. What percent or fraction of air is pure oxygen? What comprises most of the rest?

4. Why does mixing air with gas before ignition give a hotter flame?

5. In glassblowing, the flame temperatures must be even hotter than those that can be obtained with your burner. Using the same fuel (methane), suggest how hotter flame temperatures are reached.

6. Have you read over the safety and technique procedures, and do you agree to try your best to follow them?

Think, Speculate, Reflect and Ponder

7. What happens to the waste products from the Bunsen burner's flame?

8. Which one of the waste products from the Bunsen burner's flame contributes to a global environmental problem? What is the name commonly given to this problem? What kinds of fuels contribute to this problem? Give some examples of these fuels?

Experiment 2

Going Metric
With the Rest of the World

Then down with every "metric" scheme
Taught by the foreign school,
We'll worship still our Father's God
And keep our Father's "rule"!

A perfect inch, a perfect pint,
The Anglo's honest pound,
Shall hold their place upon the earth,
Till Time's last trump shall sound!

Samples From Home

1. Bring an empty can or bottle that has the volume listed on it. (For use in procedure step #2).

2. Bring ONE of the following: at least 15 mL ($^1/_2$ ounce) of a liquid (rubbing alcohol, canned fruit syrup, juice, cleaning solvent, etc.) OR an equal volume of a solid which will fit into a one inch diameter cylinder (wood, rock, metal, plastic, etc.) OR an equivalent weight of granular solid which will NOT dissolve in water. (For use in procedure step #5.)

3. Bring one small unopened food package on which a net weight is given such as a bag of peanuts, candy bar, etc. (For use in procedure step #6.)

Objectives

Various pieces of laboratory apparatus for measuring lengths, volumes, and weights will be examined. The "sizes" of familiar units of measurement will be determined in metric units and English/metric conversion factors calculated from experimental data. The experimental measurement of density will be explored and the truth-in-packaging of a consumer product determined.

Background

In the year 2010, a person goes in for a medical check and weighs in at 55 with a height of 167. Does that suggest some radical mutation had taken place in our evolution during the beginning of this century? It could happen, but not from chemical mutagens or nuclear radiation. Rather, the cause is something much less drastic although still greatly affecting industrial output, personal habits, and the world economy. All this simply marks the official coming of the metric system to America sometime in the near future.

Did you know that when the U.S. Navy ordered some cannon balls in the 19th century, in addition to stating the "precise" diameter of the balls in inches, it also included three barleycorns to use as a reference standard for the "precise" inch? Just as the size of barleycorns obviously varies from plant to plant and kernel to kernel, so did the "standard" inch. Other equally ludicrous standards exist for other units in the English system which, although considerably refined nowadays, still fall far short of the fantastic accuracy required by science—especially, for example, in the space program.

> ### 30 grams of prevention
> ### is worth
> ### 0.454 kilograms of cure

Wouldn't it be neat to have, say, a unit of length that had an almost infinite degree of exactness, was the same unit used the world over, and which could be converted to smaller or larger units simply by moving the decimal to the right or left. For instance, think how practical it would be if 1 foot = 10 inches = 0.1 yards. Well that is what the metric system is all about. The system itself has been around for a long time (since 1791), but only now is the United States seriously considering joining the world metric community. Why?

You don't need a chemistry course to tell you that people and their ways and ideas are very resistant to change; like the mass of an object relates to its inertia, so we can have an inertia of thought or action. Some might call it tradition. When dealing with a single individual, a degree of physical arm-bending could make one break with tradition. With nations of individuals, however, a more political form of arm bending can often be equally effective where all else has failed—that of economics. What could be more English than England, but that country has essentially completed the enormous job of changing over to the metric system—not just currency, but every nickelodeon, screw, and imperial gallon has received a new metric facelift.

Is there no safe place for tradition and the English system? But of course it is the United States which, in spite of much talk, still remains as the only major nation not officially committed to "going metric." (Burma and Liberia may be the only other nations still officially retaining the English system.) Nevertheless, we see the economics of world trade finally forcing countries which are the last bastions of the English measurement system to join the rest of the metric world.

Perhaps the most disastrous example of speaking different measurement "languages" occurred on September 23, 1999. Scientists that designed the NASA Climate Orbiter unwittingly combined a mismatch of English and metric units which caused the (fortunately unmanned) $125 million spacecraft to crash on the planet Mars.

Apart from measurement units themselves, *density* is a concept which is also explored in this experiment. It is not unique to the metric system, but it will be expressed only in metric units in your chemistry course. You would say that lead is heavier than aluminum, right? But you don't really mean just heavier in pounds or grams, because however much weight of lead you wish to imagine, a pile of aluminum of equal weight can be made. What you really mean subconsciously is that given equal volumes of both lead and aluminum, the lead will weigh more. This is what *density* is all about, and its units do indeed refer to a weight (to be accurate scientists say mass), but it is expressed as a weight per an equal reference volume. Metrically speaking, one cubic centimeter (1 cm³) of lead weighs 11.4 grams, while 1 cm³ of aluminum weighs only 2.7 grams. Scientifically, the *density* of lead is 11.4 grams per cubic centimeter (11.4 g/cm³), and the *density* of aluminum is 2.7 g/cm³. Unitwise, note that to obtain the density of a substance, we must divide weight by volume units (or divide volume units into weight units, if you prefer).

The densities of different substances represent a characteristic physical property identifying that substance, just like a melting point and a boiling point. All samples which are chemically pure water, for example, melt at 0 °C, boil at 100 °C under one atmosphere pressure, and have the same density which is 1.0 g/cm³ (or 1.0 g/mL since 1 cm³ = 1 mL).

Procedure

1. *Length:* Choose any object or accurately measurable distance in the lab between one and three feet long (for example the width of a lab drawer or counter top). Use a combination meter/yard stick and measure the length in both metric and English units to the degree of accuracy specified on the report sheet.

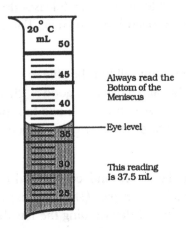

Always read the Bottom of the Meniscus

Eye level

This reading Is 37.5 mL

2. *Volume:* Fill your container from home with water to the level of the original liquid and then pour the liquid into a graduated cylinder to determine its total metric volume. Use the larger cylinders available if you have a lot of liquid to measure.

Remember when reading volumes in your cylinder to place your eye directly opposite the water level and then take the reading corresponding to the *bottom* of the concave liquid surface (called the meniscus). Examine the markings on the graduate carefully and decide how many milliliters each single division is worth.

Figure 2.1. Reading the meniscus.

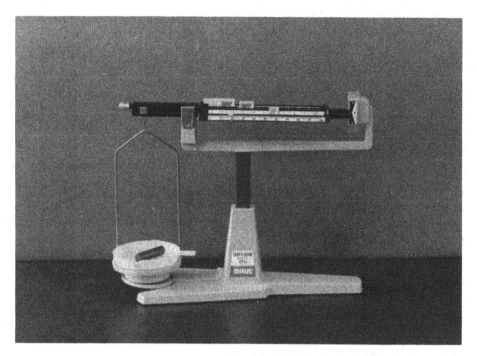

Figure 2.2. Weighing metal bar on a triple beam balance.

3. *Weight:* You will be given a cylindrical metal bar with a symbol stamped on it representing the chemical element of which the bar is composed. If no symbol is present, your lab instructor will give it to you. Using the triple beam balance as demonstrated by your lab instructor, obtain the weight of your metal bar. Always remember to <u>zero</u> the balance before you start, and to read the balance as accurately as warranted; that is, get all the weight numbers to which you are entitled. This normally will mean an accuracy to the nearest hundredth (0.01) of a gram—two decimal places.

4. *Density of a metal bar:* In order to calculate the density of your metal bar, you will also need to know its volume. This can be easily measured by determining what volume of water the bar displaces. Add water to your graduated cylinder (a 25 mL size is best) until about half full and read the liquid level to nearest 0.1 mL. Then, inclining the graduate, slide in your bar (please—don't BOMB in your metal—too many graduated cylinder bottoms have been lost that way). The increase in volume will then equal the volume of your metal.

This metal bar is really a *known* substance having a *known* density. You will thus be able to compare your *experimental* density with the *accepted* density of your bar and judge how good a "chemist" you have been. When such accepted reference data is available, scientists often like to report their results in a way that compares the actual measurement error to the total size (or value) of the object being examined.

For example, a measurement error of one yard may sound very large—and indeed it would be huge had you been measuring the width of a one yard wide bench top in part (1) of this experiment. This same one yard error, however, might be relatively minor if determining the length of a one hundred yard football field.

In order to put the size of the error into proper perspective, scientists calculate what is called the percent error, using a simple equation similar to that for calculating percent itself:

standard percent calculation $= \frac{\text{part}}{\text{total}} \times 100 = \%$

percent error $= \frac{\text{actual size of error}}{\text{accepted size (value)}} = \frac{\text{difference between accepted \& your value}}{\text{accepted value (handbook)}} \times 100 = \%$ error

Using the above equation, our one yard error in the lab bench measurement would give a 100% error, whereas the same one yard error for the football field results in only a 1% error!

Look up the accepted literature value for the density of your metal bar in a reference book like *The Merck Index* in either of the tables on the next two pages. (HINT: In many references the abbreviation d refers to density, and d^{20} means density at 20 °C.) You should repeat your measurements if your percent error exceeds 10%.

5. *Density of an unknown:* Determine the density of your unknown sample from home by measuring the mass and volume. A simple method for determining liquid densities is to weigh a graduated cylinder empty, then fill it up to a convenient level with your unknown liquid and reweigh. The difference between the two weights equals the net weight of your unknown while its volume is read directly from the graduate.

Calculate the density by dividing the mass (weight) by the volume. Liquid densities can also be determined with an hydrom-

eter if it is available. See your lab instructor for this. If you know or can estimate the possible identity of your sample from home, look up its accepted density with help from the tables on the following two pages.

6. *Truth-in-packaging:* Complete the data called for on the report sheet for your "goodies."

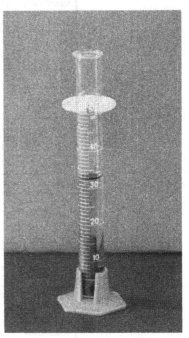

Figure 2.3. Determining the volume of a metal bar by water displacement.

Densities of Some Selected Solids

Substance	Grams per cu. cm	Pounds per cu. ft.	Substance	Grams per cu. cm	Pounds per cu. ft.
Agate	2.5-2.7	156-168	Diamond	3.01-3 52	188-220
Alabaster, carbonate	2.69-2.78	168-173	Dolomite	2.84	177
Alabaster, sulfate	2.26-2.32	141-145	Ebonite	1.15	72
Albite	2.62-2.65	163-165	Emery	4.0	250
Amber	1.06-1.11	66-69	Epidote	3.25-3.50	203-218
Amphiboles	2.9-3.2	180-200	Feldspar	2.55-2.75	159-172
Anorthite	2.74-2.76	171-172	Flint	2.63	164
Asbestos	2.0-2.8	125-175	Fluorite	3.18	198
Asbestos slate	1.8	112	Galena	7.3-7.6	460-470
Asphalt	1.1-1.5	69-94	Gamboge	1.2	75
Basalt	2.4-3.1	150-190	Garnet	3.15-4.3	197-268
Beeswax	0.96-0.97	60-61	Gas carbon	1.88	117
Beryl	2.69-2.7	168-169	Gelatin	1.27	79
Biotite	2.7-3.1	170-190	Glass, common	2.4-2.8	150-175
Bone	1.7-2.0	106-125	Glass, flint	2.9-5.9	180-370
Brick	1 4-2.2	87-137	Glue	1.27	79
Butter	0.86-0.87	53-54	Granite	2.64-2.76	165-172
Calamine	4.1-4.5	255-280	Gum arabic	1.3-1.4	81-87
Calcspar	2.6-2.8	162-175	Gypsum	2.31-2.72	144-145
Camphor	0.99	62	Hematite	4.9-5.3	306-330
Caoutchouc	0.92-0.99	57-62	Hornblend	3.0	187
Cardboard	O.69	43	Ice	0.917	57.2
Celluloid	1. 4	87	Ivory	1.83-1.92	114-120
Cement, set	2.7-3.0	170-190	Leather, dry	0.86	54
Chalk	1.9-2.8	118-175	Lime, slaked	1.3-1.4	81-87
Charcoal, oak	0.57	35	Limestone	2.68-2.76	167-171
Charcoal, pine	0.28—0.44	18—28	Linoleum	1.18	74
Cinnabar	8.12	507	Magnetite	4.9-5.2	306-324
Clay	1.8-2.6	112-162	Malachite	3.74.1	231-256
Coal, anthracite	1. 4-1. 8	87-112	Marble	2.6-2.84	160-177
Coal, bituminous	1.2-1.5	75-94	Meerschaum	0.99-1.28	62-80
Cocoa butter	0.89-0.91	56-57	Mica	2.6-3.2	165-200
Coke	1.0-1.7	62-105	Muscovite	2.76-3.00	172-187
Copal	1.04-1.14	65-71	Ochre	3.5	218
Cork	0.22-0.26	14-16	Opal	2.2	137
Cork linoleum	0.54	34	Paper	0.7-1.15	44-72
Corundum	3.9-4.0	245-250	Paraffin	0.87-0.91	54-57

Densities of Some Selected Liquids

Liquid	Grams per cm^3	Pounds per $ft.^3$	Temp $^{\circ}C$
Acetone	0.792	49.4	20
Alcohol, ethyl	0.791	49.4	20
Alcohol, methyl	0.810	50.5	0
Benzene	0.899	56.1	0
Carbolic acid	0.950-0.965	59.2-60.2	15
Carbon disulfide	1.293	80.7	0
Carbon tetrachloride	1.595	99.6	20
Chloroform	1.489	93.0	20
Ether	0.736	45.9	0
Gasoline	0.66-0.69	41.0-43.0	
Glycerine	1.260	78.6	0
Kerosene	0.82	51.2	
Mercury	13.6	849.0	
Milk	1.028-1.035	64.2-64.6	
Naphtha, petroleum ether	0.665	41.5	15
Naphtha, wood	0.848-0.810	52.9-50.5	0
Oil, castor	0.969	60.5	15
Oil, cocoanut	0.925	57.7	15
Oil, cotton seed	0.926	57.8	16
Oil, creosote	1.040-1.100	64.9-68.9	15
Oil, linseed, boiled	0.942	58.8	15
Oil, olive	0.918	57.3	15
Sea water	1.025	63.99	15
Turpentine (spirits)	0.87	54.3	
Water	1.00	62.43	4

Going Metric

Date _____   **Section number** _____   **Name** _____

1. Length of object or distance

    (a) To nearest 0.1 centimeter .... _____cm.

    (b) To nearest 0.1 inches .... _____in.

    (c) Calculated number of centimeters in 1 inch
    (divide line 1(a) by line 1(b)) = ... _____cm/in.

2. Volume of container

    (a) Experimentally measured to nearest 1 milliliter .... _____mL.

    (b) Metric volume stamped on the container itself .... _____mL.
    (If volume only appears in ounces, multiply
    number of ounces by 30 to convert into milliliters)

    (c) Based on your data in (a) and (b) above,
    does this container give you your "volumes worth"?

3. Weight of known metal bar to nearest 0.01 g .... _____g.

4. Density of a known

    Final volume of water with bar in cylinder
    (estimate to 0.1 mL) .... _____mL.

    Initial volume of water in cylinder without bar
    (estimate to 0.1 mL) .... _____mL.

    (a) Volume of metal bar
    (Subtract initial volume from final water volume) .... _____mL.

    (b) Calculated density of bar
    (line 3 divided by line 4(a)) = ... _____g/mL.
    (Units of g/cm^3 equivalent since 1 mL = 1 cm^3)

    (c) Identity of metal (from symbol on bar) _____.

(d) Accepted density of this metal (from *The Merck Index*) .... _____g/cm^3.

(e) Percent error in density measurement .... _____%.

5. Density of an unknown

    Nature of sample _____.

    Weight of sample .... _____g.

    Volume of sample .... _____mL (cm^3).

    (a) Calculated density of unknown
    (mass divided by volume) .... _____g/cm^3.

    (b) Does this value for density seem reasonable?

(Look up densities of similar substances in the density tables on the previous pages. All water based (aqueous) solutions will have a density very close to that for pure water—1.00 gram per milliliter.)

6. Truth-in-packaging:

    Brand and nature of food product _____.

    Weight of package and contents (to nearest 0.01 g) .... _____g.

    Weight of package minus contents .... _____g.

    (a) Net weight of contents (from your data above) .... _____g.
    (Subtract *empty package* weight from *package + contents* weight.)

    (b) Stated metric net weight from label .... _____g.
    (If label weight in 6(b) is only given in ounces, you will have to
    multiply the number of ounces by 28.3 to convert into grams)

    (c) Does this package truthfully offer you your "grams worth" of goodies?

7. Conclusions and comments on experiment.

Going Metric

Date _____ **Section number** _____ **Name** _____

1. Using the same basic procedure of liquid displacement followed in this experiment, describe how a volume of 10 grams of sugar might be determined experimentally in the lab. (Hint: a <u>water</u> displacement method could not be used since when solids dissolve in a liquid, the volumes are not additive.)

2. Polychlorinated biphenyls (PCBs) are a type of chlorinated hydrocarbon having densities around 1.3–1.4 g/cm^3. They are not an insecticide (like DDT) but were once used in electrical equipment, hydraulic fluid and carbonless carbon paper; however, their potential danger to the environment appears to exceed that of DDT. Not only are they very resistant to biodegradation, if spilled into water PCBs pose a much more difficult clean up problem than, say, oil. Explain why this is so given the fact that the density of water is 1.0 g/cm^3.

3. What would be the most likely weight/height measurement units for the person going in for a medical checkup? (See the background discussion at beginning of this experiment.)

4. Write the English equivalent of these metricized expressions appearing earlier:

(a) 30 g of prevention is worth 0.454 kg of cure;

(b) Give them 2.54 cm and they'll take 1.61 km.

Think, Speculate, Reflect and Ponder

5. Just what is the advantage of having everyone on the planet using the same units of measurements?

6. Why do you think that the United States has been so slow in adopting the metric system?

Experiment 3

Recycling Aluminum Chemically

Samples From Home

1. Bring aluminum from a used consumer product (beverage can, TV dinner plate, or pie tin).

2. Bring an object with a painted or varnished surface for testing the effectiveness of your synthesized paint remover (a common yellow pencil is fine).

Objectives

The recyclability of "used" aluminum atoms will be illustrated by chemically converting waste aluminum packaging into a new compound and isolating the final product using the techniques of suction filtration and drying. This compound will be added to an organic solvent and the effectiveness of the resulting solution as a paint/varnish remover will be tested.

Background

Much is heard of recycling nowadays. Since our natural resources on spaceship Earth are decidedly finite, we cannot indefinitely continue as a throwaway society—currently 4.5 lbs of waste per day for every American (see Table 3.1). We must develop a recycle mentality, and it should be encouraged and rewarded by economic incentives and public respect and appreciation. Movement towards large public support for recycling, however, will demand changes in our basic lifestyles. Such changes often require crisis or revolution, and it is difficult to impress people with a crisis *before* it happens.

Municipal Waste in the United States		
Year	Waste generated per person per day	Percent of total waste recycled
1960	2.7 pounds	- - - - -
1980	3.7 pounds	10%
1990	- - - - -	16%
2000	4.5 pounds	30%

Table 3.1. Municipal Waste.

The size of most European countries, their close proximity to each other, and their centralized economies has necessitated recycling programs there for decades. The United States is clearly far behind the rest of the developed countries/Western World when it comes to ecological use of energy and resources. Those systems' "socialized" treatment of these problems has brought much more success than our "free market" ways of letting the price of the recycled material control whether or not something can be recycled. Our system's failure lies in the unmetered cost of a polluted environment—a cost that is seldom factored into the price of virgin glass, paper or plastic.

Curbside recycling has undoubtedly helped to increase the recyling rate in the U.S. as seen in Table 3.1. And recycling rates would certainly be stimulated if the costs of recycling goods were significantly cheaper than obtaining them from virgin (or at least natural) resources. In spite of all the hoopla, unfavorable economics still makes it difficult for recycling to compete on a mass scale. There are pressures to continue special tax breaks for companies extracting natural resources from the earth, as well as reduced rate transportation costs for "natural" resources (as opposed to "unnatural", or used resources—i.e., recyclables).

Paper, #1 and #2 plastic, cardboard, paper, newsprint, aluminum, steel ("tin") cans, yard debris, glass, magazines—most or all of these *can* be recycled by consumers even if they do not have curbside recycling available, but they must be willing to persevere and find centers accepting these goods. Some states have passed beverage container deposit laws; recycle programs are being established in the work place; and cities are making efforts towards recycling programs which

could help their "Where do we put our garbage?" crisis. (New York City, however, recently ceased curbside recycling!) But the momentum *is* building—with a little help from crisis economics and the desire for a less polluted environment.

Without curbside recycling, consumers find that oft times different centers must be sought out for different goods and the nearest ones may lie very far away indeed. Consumers must furthermore be willing to accept nothing or little more than "gas money" for their labors. Therefore, at this point in time, the recycling of most things for most people must often be more an act of conscience and concern for their future and the environment, rather than the thought of any monetary gain. This is especially true since increased recycling activity has reduced prices paid to consumers (especially newsprint)—the law of supply and demand! And we need to ponder that, with less than 6% of the world's population, those in the United States consume 30 - 40% of the world's resources. Think about that!

But one castaway that is indeed worthwhile to recycle is *aluminum*. Depending upon location and type of aluminum, consumers have been typically paid 15 to 50 cents a pound and more. This metal has a relatively high initial cost, is very light weight and, most importantly, requires a great amount of energy in order to obtain it in a pure state from its natural ores. (The recycling of aluminum consumes only about 5% as much energy as that needed to liberate the metal from its natural ores.) The combination of these three factors makes it possible for *recycled* aluminum to compete economically with *natural* aluminum from ores.

"A rose is a rose is a rose" is a saying familiar to many. Likewise, *"An aluminum atom is an aluminum atom is an aluminum atom."* Used, unused, heated, burned, stomped, or pounded—all aluminum atoms must remain aluminum. Thus the secondhand "used atom lot" can be a place for real bargain hunters since the secondhand product is identical to the original! Where else besides a "used atom mart" can one find such a deal?

You begin by *dissolving* (actually, chemically reacting) your sample of aluminum with hydrochloric acid to produce a new chemical compound called aluminum chloride:

$$2Al \quad + \quad 6HCl \quad \longrightarrow \quad 2AlCl_3 \quad + \quad 3H_2$$

| aluminum metal | hydrochloric acid (water solution) | aluminum chloride (soluble) | hydrogen gas (bubbles) |

The amount of aluminum metal specified in the directions is twice as much as can react completely with the acid, so do not expect all the metal to disappear. This reaction chemically changes the aluminum metal into the compound aluminum chloride. The aluminum in aluminum chloride will then be in the form of positive aluminum ions, and when mixed with appropriate negative ions, can combine with them to yield an insoluble precipitate. The particular negative ions chosen are those which will form special kinds of aluminum salts having the ability to thicken many organic liquids so much that they will not even flow (form a gel). These substances, effective at low concentrations, are aluminum salts of certain organic acids, and they find a multitude of uses spanning the full spectrum, from the manufacture of jellied solutions for removing *dead paint* to the production of jellied napalm for removing *live flesh*. Needless to say, we will only examine the former application in this experiment!

The particular organic acids used to make these aluminum compounds in this experiment will be those found in Ivory soap. Actually, the sodium salts of these organic acids are what comprise the soap, whose principle ingredient is the sodium salt of stearic acid—sodium stearate. But Ivory soap is not pure sodium stearate. Since 1882, the label has proclaimed Ivory to be 99 and $^{44}/_{100}$ % pure, purportedly due to an analysis by a chemist hired by Harley Proctor—son of the founder of the Proctor & Gamble company. Ivory soap was thus found to contain 99 $^{44}/_{100}$ % "fatty acids and their alkali salts", the remaining $^{56}/_{100}$ being "foreign and unnecessary substances." So voilà, the ad campaign was born!

The chemical reaction can proceed in several steps giving several different, albeit similar, products. Thus the product you isolate will not be one pure compound, but a mixture of several compounds. The process is not unlike that which occurs between soap and certain other ions (calcium and magnesium) which are found in hard water, giving us the once familiar "bathtub ring" (see Experiment 10: *Why is Water Harder than Iron*). Both of the following reactions probably occur to some extent in your preparation:

$$H_2O \; + \; AlCl_3 \; + \; 2Na\,C_{18}H_{35}O_2 \; \longrightarrow \; Al\,(C_{18}H_{35}O_2)_2\,OH \; + \; 2NaCl \; + \; HCl$$

$$2H_2O \; + \; AlCl_3 \; + \; Na\,C_{18}H_{35}O_2 \; \longrightarrow \; Al\,(C_{18}H_{35}O_2)\,(OH)_2 \; + \; NaCl \; + \; 2HCl$$

water	aluminum chloride (soluble)	sodium stearate (soluble)		aluminum stearate complex salt (insoluble)	sodium chloride (soluble)	hydrogen chloride (soluble)

Warming the solution containing the precipitate enhances its thickening power and helps coagulate it into small particles which can be lifted from the surface. The acetone wash removes adhering water and permits rapid drying of the precipitate so that it can be quickly tested as an organic solvent thickening agent. You might like to try squirting a little acetone (common nail polish remover ingredient) on your hand to observe how quickly it evaporates and cools your skin.

A mixture of organic solvents is then added to your aluminum salt to form the thick liquid or gel necessary for a good paint remover. It is actually these solvents themselves (mainly methylene chloride) that work their way underneath the paint film and "lift" it off. But if these solvents were not made into a jelly consistency, they would quickly run off of the surface and evaporate before they had time to do their paint removing job.

Many commercial paint removers have, in addition to methylene chloride as the main ingredient, other chemicals to enhance their solvent power and thickening ability, such as aromatics, alcohols, etc. The particular formulation used in this experiment is typical and can be found in Bennett's *The Chemical Formulary*, a multivolume encyclopedic reference which offers how-to-make and ingredient information on a multitude of products from <u>a</u>dhesives to <u>w</u>indow cleaners. (You might like to browse through some of the *Formulary* volumes next time you are in a library.)

Aluminum ions can also be used to make gels to thicken water. When mixed with hydroxide ions (OH^{-1}) found in alkaline (basic) solutions, the following reaction occurs:

$$\text{Al}^{+3} \quad + \quad 3\text{OH}^{-1} \longrightarrow \text{Al}_2(\text{OH})_3$$

aluminum   hydroxide   aluminum hydroxide
ions (soluble)   ions (soluble)   (insoluble jelly)

But the jelly is very touchy—too much (or too little) hydroxide will make the gel disappear. Particles of this cheap aluminum gel have been used to clarify lakes and reservoirs by sticking suspended sediments to its surface as the gel falls to the bottom.

Procedure

1. Weigh out about 1 gram of aluminum from your food/beverage container and place into a 100 mL beaker (exactness not necessary since an excess of metal is used). If you choose an aluminum beverage can, use the sides.

Aluminum parts often have an obvious coating of paint or lacquer on them which will retard its reaction with acid. Scrape off ¾ of the coating with a knife or razor blade. Cut up into small pieces if necessary, or press down so that your sample lies flat on the bottom of the beaker and can be completely covered by the acid.

Figure 3.1. Scraping paint from can.

 Place your beaker plus sample in a hood and, using a 10 mL graduate, add 10 mL of dilute (6M) HCl. Depending upon your particular aluminum sample, a vigorous reaction will occur immediately or within 20–30 seconds as shown on the next page. Swirl (do not stir) the beaker contents periodically until the reaction has ceased or becomes very sluggish (about 10 minutes). Record observations on the report sheet.

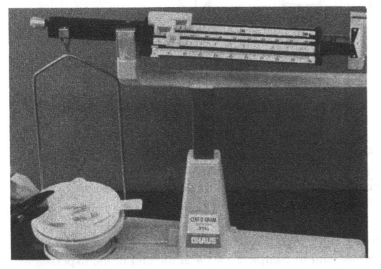

Figure 3.2. Cutting and weighing aluminum sample.

Figure 3.3. Adding hydrochloric acid to aluminum.

Figure 3.4. Aluminum reacts vigorously with acid.

2. Take reaction beaker back to your desk station and add 20 mL of water to it. Mix well with a stirring rod and gravity filter the beaker contents by pouring the liquid through a 12.5 cm filter paper cone fitted inside of your funnel. Catch the liquid in a 125 mL Erlenmeyer flask. This clear liquid (called *filtrate*) represents your **recycled aluminum stock solution.** As soon as you have collected 5 mL of filtrate (use a 10mL graduate to measure), you can proceed with the next step— but let the filtration continue to obtain more filtrate necessary for step 4. **Repeat: do NOT forget to SAVE this liquid.**

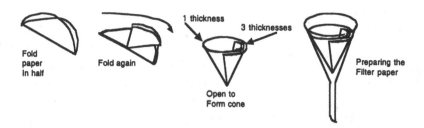

Figure 3.5. How to fold a filter paper to form a cone.

(a) Pour 200 mL of a 0.5% soap solution into a 400 mL beaker. (Your lab instructor will advise you if you are to get a special 400 mL beaker from the stockroom just for use in this experiment. If so, you need to return this beaker at the end of the experiment.)

While stirring vigorously with a spatula or glass rod, add to this beaker containing the soap solution 5 mL of your aluminum stock solution. Use a 10 mL graduate for this addition. Note your observations. Again, don't forget to SAVE the aluminum stock solution left over in your Erlenmeyer flask to recoup any "goofs" and to complete Part 4 at the end of the experiment.

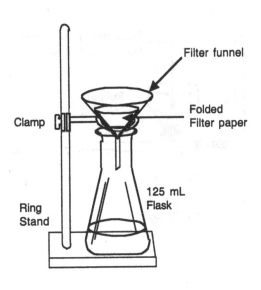

Figure 3.6. A gravity filtration set-up.

Figure 3.7. Filtering aluminum reaction.

Place the beaker of solution containing precipitate onto a wire gauze mounted on a ring and heat as rapidly as possible with a Bunsen burner placed directly underneath. As the liquid heats up, use the blunt end of a scopula to skim off the granular precipitate which rises to the surface. But WATCH THE LIQUID LIKE A HAWK and turn off the heat before it boils over.

Transfer as much as feasible of this solid (maybe 90%) directly to another 400 mL beaker about ½ full of cold water and break up any lumps with the scoopula. Occasionally, a *gum ball* instead of a granular precipitate may be obtained. If you are the unlucky person or group, tell your lab instructor about your misfortune. You have the option of starting over or just teaming up with another group.

Figure 3.8. Heating aluminum solution with soap.

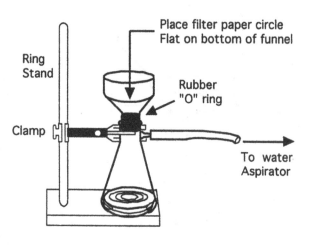

Figure 3.9. Skimming off aluminum/soap precipitate.

Figure 3.10. Büchner funnel and filter flask for a suction filtration.

 You will now filter the solid from the cold water using the technique of *suction filtration.* Your lab instructor will demonstrate this procedure, but the essential steps are these:

> Obtain a 250 mL or 500 mL suction flask, 7 cm diameter Büchner funnel, rubber "O" ring and filter paper (5.5 cm paper circles), to fit inside the funnel.

> Clamp your flask to a ring stand to prevent overturning.

> Hook up the suction flask to the water aspirator with the heavy wall rubber tubing provided.

> Place the "O" ring onto top of the flask with the flat side facing down. Center the hole over the flask.

> Insert the Büchner funnel and drop the filter paper in place so all the holes are covered.

> Wet the filter paper with water (sprinkling with your hand is OK).

> Turn on the cold water FULL and press down on the funnel by laying the palm of your hand flat across the funnel mouth until you can feel the suction take hold.

 Now you are ready to proceed with the filtration of your sample. With the suction applied as described, pour your beaker contents into the funnel as rapidly as the liquid is sucked through. Use your scoopula to transfer any remaining bits of solid. (Your lab instructor can assist you if the filtration proceeds too slowly due to a plugged filter paper.)

Stop the suction by turning off the water. Add approximately 10 mL of acetone using a 10 mL graduate directly to the white solid in your funnel (CAUTION: *Flammable liquid. Make sure that there are no flames in the laboratory.*) Gently—so as not to tear the filter paper—mix the precipitate with acetone using the blunt end of a solid glass rod. Turn on the water again and hold down the sides of the funnel until the suction takes hold and pulls all the acetone through the paper. Maintain the suction for two minutes. Discard filtrate and save the white solid left in the Büchner funnel. This solid is the aluminum soap precipitate.

Weigh a 50 mL beaker to the nearest 0.01 g and record weight on the report sheet. (Your lab instructor may direct you to get a special 50 mL beaker for use just in this experiment which you should afterwards return.) Dump/scrape your solid aluminum soap precipitate left in funnel into this 50 mL beaker.

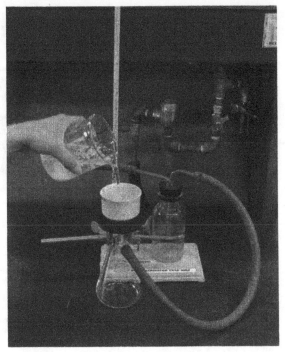

Figure 3.11. Filtering precipitate using a suction apparatus.

Place the beaker into a drying oven for 5 minutes at 150 °C. Remove the beaker from the oven using tongs or paper toweling to prevent burning your hands on the hot beaker and let the beaker cool to room temperature. The solid should be bone dry and have no acetone smell whatsoever.

Figure 3.12. Scraping precipitate into a beaker.

Figure 3.13. Drying precipitate in an oven.

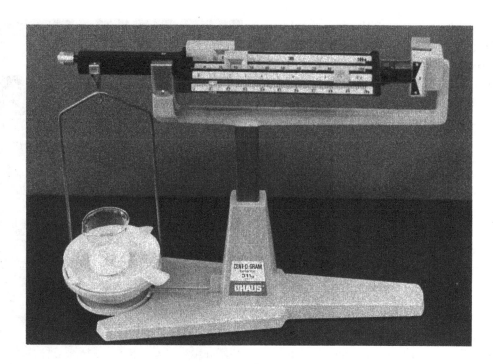

Figure 3.14. Weighing aluminum/soap precipitate.

(b) Note the texture and appearance of this solid (color, crystal form if any, etc.).

(c) and (d) Reweigh the beaker plus contents to the nearest 0.01 gram and report the net increase in weight as the weight of precipitate on the report sheet. Also calculate the percent yield using the formula given on the report sheet.

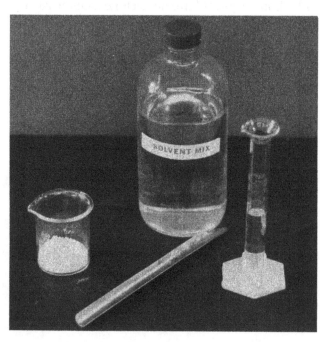

Figure 3.15. Ready to add solvent mixture to powdered aluminum soap.

3. Thoroughly powder your white solid aluminum soap precipitate by mashing with a scoopula. Then pour 5 mL of the solvent mixture solution (look for the bottle with this name on the label) into a DRY 10 mL graduate. If necessary, dry by rinsing with a little acetone and allow to drain.

With the aid of a capillary dropping pipette (long nose medicine dropper), add the solvent mixture solution in your 10mL graduate to the white powder in your 50 mL beaker in roughly ½ to 1 mL portions, stirring and mashing up the powder/liquid mixture after each portion is added. STOP adding more solvent when the beaker contents attain a honey consistency. (Expect each 0.1 g. of solid to require about 1 mL of solvent; e.g., 0.5 g solid needs 5 mL solvent.)

(a) Record the approximate volume of solvent added.

(b) Note the appearance of the gel.

(c) Spread your solvent mixture/aluminum soap complex salt "solution" from step 3 onto a painted or varnished surface and allow it to stand 5–10 minutes while doing Step 4. A common yellow painted pencil works fine. Evaluate the effectiveness of your paint remover.

Figure 3.16. Testing paint removing ability of gel on a pencil.

4. Add dilute ammonium hydroxide [$6M$ NH_4OH] to the unused remainder of the aluminum stock solution in the 125 mL Erlenmeyer flask from part 2. Squirt the ammonium hydroxide into the flask with a medicine dropper while continuously swirling. When the proper amount (not more, not less) of NH_4OH has been added, aluminum hydroxide will have formed and the "liquid" at this point should be so thick that it will not pour. This is an example of how water also can be made into a gel using an aluminum compound.

Record your observations from the nature of $Al(OH)_3$ on the report sheet.

Recycling Aluminum Chemically

Date _____   **Section number** _____   **Name** _____

Source of your aluminum sample_____

1. Observations from adding acid to aluminum metal _____

2. Aluminum solution + soap
    (a) Observations upon addition _____

    (b) Texture and appearance of dried precipitate _____

        weight of beaker + dried Al soap ...._____g.

        weight of empty beaker ...._____g.

    (c) net weight of Al soap ...._____g.

    (d) Per cent yield of Al soap
    (grams in line 2(c) is what % of the maximum possible yield of about 1.0 g.)

    (divide line 2(c) by 1.0 g and then multiply by 100) ...._____%.

3. Formation of a gel
    (a) Volume of solvent mixture added to powder ...._____mL.

    (b) Appearance/nature of the gel _____

    (c) Comment on the effectiveness of above "solution" as a paint remover.

        _____

4. What is the observed nature of $Al(OH)_3$?

    _____

5. Comments and conclusions on experiment.

Recycling Aluminum Chemically

Date _____   **Section number** _____   **Name** _____

1. Does the aluminum really "dissolve" in the hydrochloric acid? Explain for credit. (Hint/help: would you get back aluminum metal if you evaporated off the liquid acid?)

2. Why does an organic thickening agent enhance the effectiveness of organic solvents as paint removers?

3. Check *The Merck Index* for two other uses for aluminum stearate.

   (a)

   (b)

Think, Speculate, Reflect and Ponder

4. Why doesn't the United States just pass a law requiring that no more virgin aluminum be used in products sold in the United States?

5. Considering the law of supply and demand, what will have to happen to make the price paid for recycled products rise? Can you play a part in this in any way?

6. What uses for aluminum are depicted in the sketch appearing at the beginning of this experiment? (Hint: the gems depicted are rubies.)

Experiment 4

Radioactivity

Sample From Home

No samples from home are required for this experiment.

Objectives

This thought experiment will give you an idea of your annual dosage of radiation and may give you a few things to think about during the ensuing debate by the lab class on the generation of electricity using nuclear power.

Background

Radioactive decay is as natural as the universe itself. As a matter of fact, the majority of the known nuclides (or isotopes—particular arrangements of protons and neutrons in a nucleus) *are* radioactive. There are about 2000 known nuclides and less than 15% of these are stable and non radioactive. The effects of radioactivity on human beings, however, can be quite dangerous. Radiation that is sufficiently energetic can damage human tissue by ionization, dissociation, and excitation of the molecules in the tissue. This can cause the death or extensive damage of this tissue. In addition to the outright killing of tissue cells, the damage that radiation does can take place in the genes—that part of the cell that carries the instructions for reproduction. This may lead to cancer.

The natural background radiation that we are exposed to from birth (actually from conception) comes from sources like the building materials that surround us everyday, cosmic radiation entering the earth's atmosphere from space, and the food we eat containing radioactive elements from natural sources. These factors are all very low and vary somewhat based on your location on the planet. People who live in stone buildings are, for instance, exposed to more radiation than those who live in wooden structures because stone contains more naturally radioactive atoms than wood. Likewise, people who live at high altitudes are exposed to higher levels of cosmic radiation than those who live at sea level because there is less of the protective blanket of the atmosphere between them and space.

Background radiation from natural sources has not changed and probably will not change significantly over time. It is determined by the composition of the earth, and in the case of cosmic radiation, our planets position in the solar system and the universe in relation to other stars, supernovae, and blackholes. The amount of radiation from "man-made" sources, however, has been on the increase ever since the explosion of the first atomic bombs in 1945. (The materials that are referred to here were of course not really made by human beings but were merely mined and refined from natural deposits or made by the reactions of naturally radioactive elements.) Between 1945 and 1988, 920 nuclear devices were detonated by the US government. Until the Atmospheric Test Ban Treaty was signed in 1963, nuclear weapons tests were often carried out at the surface of the earth where the radioactive products of these massive explosions contaminated the air, soil, and water. Between 1946 and 1963 the United States conducted over 130 open air nuclear tests.

Since the Atmospheric Test Ban Treaty, nuclear weapons testing has been voluntarily limited to underground explosions which are generally not vented to the atmosphere, although radiation venting has happened by accident on a number of occasions by many countries including the U.S. As recently as 1986 a poor estimation of the local rock strength during an underground U.S. test caused the venting of radiation at the Nevada Test Site operated by the U.S. Department of Energy.

Testing by the countries that used to be the U.S.S.R. ceased in the 1990s and the U.S. has, to a degree, slowed its active nuclear underground testing in response. A relatively recent example of an underground nuclear test occurred October 7th, 1994 when a "medium to large" nuclear device was detonated by The People's Republic of China, the third that they had tested in a year. Just as nuclear testing by the U.S.S.R. used to make the U.S. nervous, China's recent testing clearly upset Japan and Taiwan.

In addition to radiation releases from nuclear weapons testing, the advent of electricity generation from nuclear power has increased the background radiation by an undetermined amount. In the case of the increased radiation dosage per human being, the increased amount averaged over the entire population of a country or hemisphere or especially the entire world is vanishingly small when compared to the natural background radiation that is already present. However, for individuals who live near the site of a nuclear accident (for example, Chernobyl in the Ukraine or the Hanford nuclear reservation in Washington state) the increase in radiation dosage may be quite significantly increased. Recent disclosures of many other radiation contaminated sites in the former U.S.S.R. also suggest that the dangerous levels of radiation there are much greater than were ever imagined even by the staunchest anticommunists of the Cold War.

Finally, don't forget to consider the contribution to our radiation exposure from extremely useful medical tools such as X-rays. These also increase our radiation exposure to some degree based on the amount of X-ray procedures performed, the equipment used, and the part of the body imaged by this technique. Though this exposure to radiation is not really a natural consequence of living in our environment, it is after all a choice that we make to have these techniques used. Few people choose not to be exposed to this sort of radiation because they believe the test to be necessary. A recent article in *Consumer Reports* did note that some chiropractors insist on using whole body X-rays more than is considered safe by the mainstream medical community. It also warned readers away from chiropractors who insisted on this method of finding problems with spine and nervous system (*Consumer Reports*, 1994).

The biological effects of radiation can be measured in many different ways. One of the units of measurement is called the roentgen equivalent man (rem). One thousandth of a rem is a millirem (mrem). The table below shows a few estimates of the annual exposures of human beings to different sources of radiation measured in millirems. Note that these are just a few of a number of widely varying estimates and are not meant to reflect accurate, well-known values.

They are only provided to give you an idea of the sources of radiation exposure in your life and the possible relative contribution from each. Also note that these values are an average and may very well not reflect dosages of individuals acutely exposed because of weapons test venting, close proximity to coal fired power plants or nuclear power plants, nuclear accidents, repeated medical X-rays, etc. The last two values in the table are quoted from *National Geographic*, August, 1994 and are considered one time doses not annual ones.

Sources of Radiation	Average annual dose in millirems
Terrestrial	29
Cosmic Rays	30
Dental X-rays	14
Medical X-rays	39
Burning of Fossil Fuels	2
Fallout from weapons testing	4
"Acceptable" Single Dose at Chernobyl Accident Cleanup	25,000 (25 rem)
Acute Radiation Sickness Dose	approximately 200,000 (200 rem)

Table 4.1. Various Estimated Radiation Exposures.

Radioactivity

Date _____   Section number _____   Name _____

Class discussion of nuclear power generation in the United States

A letter to the editor in a widely read magazine flatly stated that the letter's author was tired of hearing and reading that individuals calling themselves "environmentalists" were opposed to the use of nuclear power to generate electricity in the United States. The writer said that anyone who opposed nuclear power (in the Northeast) was in effect choosing brownouts and power failures and this, he stated, was an obviously anti-environmental position since many people's lives depend on a relatively constant flow of electricity (hospitals, drug storage, food preservation, etc.). The author concluded that of all the present energy methods to supply mass power grids, nuclear power was the cleanest and safest means presently available in the United States.

To begin with, do you agree or disagree with this author's position? Why or why not? Discussion should initially be focused upon this point but can go farther afield. Areas such as alternate means of power generation, pollution factors—especially global warming, waste storage, possibility of accidents, and health concerns are all fair game for this discussion. All members of the laboratory will be expected to express an opinion on the initial question at least, but involvement and participation in this discussion will not be forced beyond the answering of the initial question. Prior reading and preparation are encouraged for this laboratory "experiment."

Radioactivity

Date _____   **Section number** _____   **Name** _____

1. The public debate about the generation of electricity from nuclear power has been raging for well over a decade. Why is this such a charged question?

Think, Speculate, Reflect, and Ponder

2. One of the solutions to the disposal of nuclear waste would be to fire these materials into space using automated rockets that fly either into deep space or into our sun. Why has this alternative been so strongly discounted *since 1986*?

3. In the 1990s there was an upsurge in the support for nuclear power in the U.S. (at least from the major utility consortia and engineering and design companies). What happened *in the summers* of the late 1980s and throughout the 1990s that helped this resurgence?

4. In the spring and fall of 1994, plutonium, a very dangerous radioactive substance used in nuclear weapons, was seized many times as it was being smuggled into Germany. In the summer of 2002 a seizure of 34.6 <u>pounds</u> of uranium was made in southeastern Turkey. Where did this nuclear material probably come from and why was it being smuggled?

5. Should the United Nations spend money from its budget (supplied by its member nations) to stop "renegade" plutonium and uranium distribution, that is, shipments of nuclear material from the producers in the former U.S.S.R. to nations that cannot make this substance themselves and yet are trying to make nuclear weapons? How could this be accomplished? What reasons are there for the member nations to do this?

6. Does the accidental release of nuclear radiation from the Nevada Test Site increase the background radiation experienced by people "downwind" of the test site?

Experiment 5

O$_2$ Content of Air

Sample From Home

No samples from home are needed for this experiment.

Objectives

The percentage of oxygen in normal atmospheric air will be determined using a simple method. This technique makes use of the catalyzed oxidation (rusting) of iron in a closed atmosphere.

Background

The oxygen content (molecular oxygen, O$_2$) of our atmosphere is something that we seldom think about, yet something that is very important. A large number of living things on this planet depend on the oxygen in the air for life. Some organisms remove oxygen from water solutions simply by the diffusion of dissolved oxygen through the membranes of the organism's cells (for example, protozoa and some kinds of worms). Larger life forms need a more specialized way to transfer oxygen into their bodies and to get waste products like carbon dioxide out of their bodies. Many marine animals have gills for this purpose to extract dissolved oxygen from the water that they live in. Some gills (like those that starfish have) are passive absorbers of oxygen and rely on water that is passing by to come in contact with the gill surface to supply the needed oxygen. Other kinds of fish (like sunfish and catfish) optimize this situation by pumping water through their gills. This process is assisted by the fish's movement through the water.

The largest animals on the planet need an even more efficient means of extracting the oxygen that they need from the air (instead of O$_2$ dissolved in water). Lungs have evolutionarily developed for this purpose. Animals with lungs get the oxygen they need directly, by breathing air into their lungs and selectively removing the oxygen from the other gases (mainly nitrogen and a small amount of carbon dioxide). Human beings, obviously, fall into this last class, and we too are extremely dependent on the oxygen content of the air we breathe.

People who go snow skiing only once a year often have a headache during their first day on the slopes. The reason for this is that most ski slopes are in the mountains at relatively high elevations (the new Japanese artificial slopes in Tokyo notwithstanding). As you move higher and higher into the atmosphere the air gets less and less dense; the heaviest air is near the bottom of the atmosphere at the surface of the earth, and the lightest air is at the top of the atmosphere. Some of the "missing" density at higher altitudes is due to a lower oxygen content in the air. Performing a relatively high impact activity like snow skiing demands a lot of oxygen. Although skiers may not consciously notice the lower oxygen content, their bodies often do and respond with a headache. People who have spent an extended time in this lower oxygen atmosphere have developed additional lung capacity to increase their ability to extract the oxygen that they need from the thinner air.

The rusting (oxidation) of iron usually takes place slowly, and in general we're happy about this. This process is, however, accelerated by moisture and acidity. The procedure in this experiment takes advantage of both of these catalysts. (Catalysts are chemical reagents that speed up a reaction but are not used up by the reaction itself. Therefore in some systems, the catalyst can be recycled.) The temperature of the reaction mixture, though not a catalyst, is important: Most reactions occur faster at higher temperatures.

The controlled oxidation of iron in this experiment is performed in a moist, slightly acidic atmosphere, and the amount of oxygen necessary for the reaction is measured by the "reverse displacement" of water. As oxygen is taken out of the air by chemical reactions on the surface of the rusting iron, the vacuum produced sucks water up into the inverted graduated cylinder to take its place. (Scientists are happier if you say that the higher air pressure outside the cylinder PUSHES the water up into the lower pressure space.) The percentage of oxygen in the air can then be calculated from the "sucked-up water volume" compared to the original air volume of the graduated cylinder. Read the experiment's step *before you start* and you will surely see the effects of rusting with your own eyes in a few minutes time.

Procedure

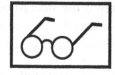

1. Determine the weight of a piece of weighing paper or plastic weighing boat. Record the weight to 0.01 g.

2. Wash your hands with soap and water and dry them. Get a piece of steel wool about ¼ as large as your fist from the supply cart or stockroom. Pinch off a single piece between two fingers and gently roll that piece into a ball in the palm of your hand. Place approximately 0.75 g of steel wool rolled into a ball on the weighing paper and record the total weight to 0.01 g. (This piece should be about as large as the end of your thumb, but the size of the ball actually depends on how tightly you roll the ball.) This weight shouldn't be any larger than 1.0 g and no less than 0.5 g. If you end up with a larger or smaller steel wool ball, pinch off a piece or add a small piece to your ball to get *approximately* 0.75 g. Don't weigh the small pieces that break off. Throw these away and return the larger pieces that you don't use to the supply cart or stockroom. After the ball is weighed, stretch it open again so that its surface area increases. This will increase the effect of the acid in the next step and its contact with O_2 later after you put it in the graduated cylinder.

3. Measure 35 mL of 0.3 M acetic acid with a graduated cylinder and pour this entire volume into a 50 mL beaker. Keep the beaker in the vent hood. *Using your tweezers*, put the steel wool weighed out in Step 2 into the acetic acid making sure it is completely covered by the acid. You may have to press down on the steel wool to submerge it.

4. Leave the steel wool in the acid for 1 minute. Remove the acid-dipped steel wool from the acetic acid solution with your tweezers, put it in a clean 50 mL beaker, and take it back to your work station or bench top. Do not wash off the steel wool after the acid dip. Unfold and "exercise" the steel wool so that you expose as much surface area as possible. Stretch it into a steel wool tube if you can. If you get the acetic acid on your hands, finish what you are doing and then put the steel wool back into its transport beaker and wash your hands with water for 30 seconds. Acetic acid is a weak acid so you don't need to rush but you do need to make sure you remove any acid you get on your hands or work bench.

Read Steps 5-7 before you begin Step 5. You now have to do a number of things correctly to get the best results. A partner can help here especially if you are working in a group.

5. As quickly as possible, remove the acid-dipped steel wool from your 50 mL beaker using your tweezers and *loosely* pack the steel wool down into the *bottom* of a clean 25 mL graduated cylinder. Try to expose as much of the steel wool to the air in the graduate as possible. Invert the graduated cylinder con-

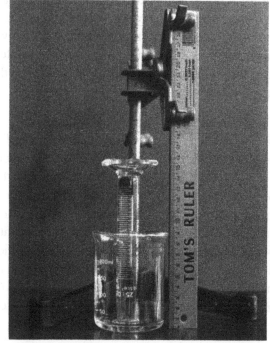

Figure 5.1. Taking the initial reading.

taining the steel wool and set it down into an *empty* 250 mL beaker. Using your wash bottle, add water to the beaker until the lip of the graduated cylinder (including the pour spout) is just covered by water. BE CAREFUL NOT TO KNOCK OVER THE INVERTED GRADUATE.

For this next step use a ruler marked in mm. Set the ruler on the bench top next to the beaker containing the inverted graduate and measure the distance from the bench top to the top of the water that you have added to the beaker. You will need to place a ruler on the bench top outside of the beaker and sight from the water level in the beaker over to the ruler's marks. Use the millimeters side of the ruler instead of the inches side. This will be a small number, probably less than 10 mm. This is your *initial water height*. (See Figure 5.1.) Record this value on the report sheet in the appropriate place.

As the oxidation (rusting that you have initiated by moistening the steel wool with water and acid) begins, the reaction will use up oxygen in the air <u>inside</u> the graduated cylinder. You will be able to see this because the water level *inside* the graduate will creep up: Oxygen will be used in the reaction and the volume of gases inside the graduate will decrease, pulling water up into the graduate from the beaker. This change in volume will be quickest at the beginning and slower later on. Why?

6. Start watching the clock as soon as you have inverted the graduate and finished measuring the initial water level. About every two minutes, carefully add water to the beaker with your wash bottle to make the water level *inside the graduated cylinder* and the water level in the beaker exactly level with each other. DO NOT LET THE WATER DROP BELOW THE GRADUATE'S SPOUT. (If the water level starts to move so radically that there is a chance that the water level *in the beaker* will drop below the spout of the inverted graduate and will allow air to get sucked into the graduate instead of water, carefully add water to the beaker with your wash bottle before the next two minute mark.) After about 10 minutes, stop making water additions with your wash bottle every two minutes and instead add water to even out the water levels only every five minutes. Again, be careful not to tip over the in-verted graduate. Continue adding water every five minutes until the reaction stops (about 20 minutes total), and the level of water inside the graduate does not rise any more.

7. When the reaction is completed (no more change in water level since your last leveling addition of water), measure the height of the present water level in the graduate *above the bench top*. In other words, get the height of the water level above the start of the ruler. Place the ruler on the bench top outside of the beaker and sight from the final water level in the beaker over to the ruler's marks (which will exactly match the height inside the graduate if you made your last addition carefully—See Figure 5.2.) Record this reading in millimeters on the re-

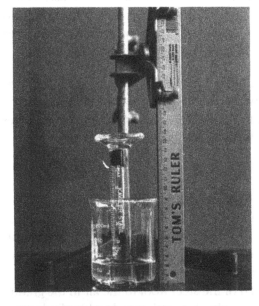

Figure 5.2. Taking the final reading.

port sheet in the appropriate place. The difference between the initial and final readings will therefore be the distance that the water has travelled up into the graduated cylinder during the whole reaction time.

After you have successfully measured the distance that the water level has travelled, remove the steel wool from the graduated cylinder and discard it in the appropriate waste container. With your ruler measure the total height of the cylinder *from the lip to the top of the base.* (See Figure 5.3.) Record this value in millimeters on the report sheet as the height of the graduated cylinder.

8. Repeat the experiment from Step 1. Use a fresh, newly weighed ball of steel wool; however, you *may* use the acetic acid left in the 50 mL beaker from before to acid rinse the steel wool. For

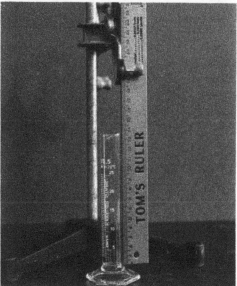

Figure 5.3. Measuring the total height of the graduate.

the second trial, if it's available get some food coloring from the supply cart or stockroom and see if adding food coloring to the water in the beaker helps you to visualize the water levels easier. Add 2 drops of food coloring to the water in the beaker after you have positioned the inverted graduated cylinder and covered the graduate's lip with water (Step 5). Record all of your data on the report sheet for the second run.

Calculate the percentage of O_2 in air for each run using the steps on the report sheet.

Report Sheet

Experiment 5

O_2 Content of Air

Date _____   Section number _____   Name _____

1. Data table
2. Calculations

	Run # 1	Run # 2
Mass of Steel Wool + Paper		
Mass of Paper		
Mass of Steel Wool		
Height of Graduated Cylinder (mm)		
Final Height of Water in Graduated Cylinder (mm)		
Initial Height of Water		
Average % Oxygen Content in Air		%

Run #1   Run #2

a) Height of graduated cylinder .... _____ mm   _____ mm.
   (from data table)

b) Calculate the change in height
   of the column of water during the experiment
   (subtract the initial ruler reading
   from the final ruler reading) .... _____ mm   _____ mm.

c) Calculate the percentage of oxygen in air
   (divide 2(b) by [2(a) minus initial height of water]
   and then multiply by 100) [all units in mm] .... _____ % O_2 _____ % O_2.

d) Record the average of these two percentages
   on the data table above. .... Average   _____ % O_2.

O$_2$ Content of Air

Date _____   **Section number** _____   **Name** _____

1. The percentage of oxygen in the air at sea level is approximately 20.6%. Calculate the percent error for your experiment:

subtract your average experimental % O$_2$ from 20.6

(accepted value - experimental = difference)

and then

divide this result by 20.6 and then multiply by 100

$$\frac{\text{difference}}{20.6} \times 100 = \text{percent error}$$

2. What is a catalyst? Are there any catalysts used in this experiment? If so, what are they?

Think, Speculate, Reflect, and Ponder

3. The oxygen content of the atmosphere has a very important effect on forest and prairie fires. What would you expect might happen to the severity and number of fires on the earth if the oxygen content of the atmosphere (near the surface) increased from approximately 20.6 % to 25 %? Conversely what would you expect might happen if the natural atmospheric oxygen content became 15 %?

4. Catalytic converters are installed in all automobiles sold in the United States. These devices are placed "in-line" in the exhaust pipe, and all gases that are exhausted by the engine pass through the catalytic converter. What might the job of this device be?

5. Could a mouse live by breathing the gases left in the graduated cylinder at the end of the experiment? Explain your answer for credit.

Spectroscopy

Introduction to Spectroscopy

Why are some molecules colored and others not? Why are some red and others green? The interaction of atoms and molecules with light is intriguing and a little complex; however, there are five laboratories in this manual that use the interaction of light with matter in an interesting manner. This section is designed to introduce you in a general way to the reasons why molecules and atoms interact with light and to introduce you to an instrument called the spectrophotometer which is designed to use light to analyze chemical samples.

Light and Color

The colors of the <u>visible</u> light spectrum run from red to blue. The old mnemonic used to memorize the colors of the visible spectrum is ROY G. BIV: red orange yellow green blue indigo violet. Lights that you see that have these colors have one thing in common: they are all the same physical phenomenon, that is, waves or particles of energy that we call light. By the way, there is an interesting set of experiments that will show light acting in one experiment as a wave and in another experiment as a particle. We will leave a description of this, the "two slit experiment," for you to find somewhere else—if you are interested. Though the results of these experiments are, in fact, quite interesting to most scientists, they are beyond our scope here and you do not need to think about how light travels—as a particle or a wave; none of the experiments that you will perform in this manual will change if you adopt one view or the other.

Light and Energy

Again, colored lights are all the same thing in one way, as we said before, but they are all different too. The *energies* of the particles or waves that make these colors are all slightly different. Therefore red light is slightly less energetic than blue light, and this has important effects when these lights interact with matter; for one, your eye interprets this difference in energies as a difference in color. When all of the colors of the visible spectrum arrive together at a surface in your eye called the retina, your brain interprets this light as being white, much like white paint can be made by blending all the colors of the rainbow into one pot; however, if the individual colors arrive separately, each is independently "felt" by your eye as that color, and the nerve signal for that color is sent to the brain by your optic nerve. Voilà! Color vision.

This energy difference is also the reason why some substances selectively absorb some colors of light and not others; however, we must discuss briefly the structure first of first atoms and then molecules before we can determine which will absorb light radiation and which will not and why.

Atomic and Molecular Structure

The structure of *atoms* includes a positively charged nucleus surrounded by negatively charged electrons. The general locations that the atomic electrons occupy around the nucleus are called atomic energy levels and have very well defined energies and relatively simple structures. This means that for a particular chemical element, like sodium, the electronic levels have only specific energies that are characteristic of that element; potassium's levels are different, as are lead's and zinc's. Individual atoms, not attached or bonded to other atoms therefore have relatively simple electronic structures since their electrons are interacting with those of any other atoms.

Molecules, on the other hand, are made up of atoms and their electrons joined together in such a way that electrons are *partially shared* by different atoms. This means that there are many more electrons in molecules than in simple atoms and that they have many more possible energy levels and structures because of this sharing. They are, therefore, electronically more complex than atoms.

Absorption of Energy

The interaction of light with molecules in solutions like water or an acid solution involves the absorption of light passing through the solution by the molecules contained therein (see Experiment 6: *Chromatography of Natural Pigments* or Experiment 16: *Warning: This Experiment My Contain Lead*). Here's the key: This absorption fundamentally requires that the electrons in the molecules must have a higher (in energy) electronic level *that they can move to* when the energy of the light is absorbed. (If no higher state is available no absorption will take place.) Therefore when the absorption of light occurs the electrons in the molecule that does the absorbing are excited to a higher energy state and the light that is absorbed does not exit the solution; it is gone—it has been absorbed by the molecule. Light is absorbed and electrons get promoted to a higher level.

The reason that some molecules will only absorb specific colors is that they have an electronic structure that has only certain energy levels available. Therefore only light of a certain color (read energy) will do. The rest of the light passing through the solution does not interact with those molecules: those molecules are transparent to those light waves. For a particular chemical solution containing a particular light absorbing molecule, only a specific color is absorbed; therefore, the rest of the light passing through the solution does just that—pass through. This means that the light exiting the solution will be missing some of the light that went in—the light that was absorbed. If no color is absorbed as light passes through, the solution will appear colorless. Pure water has no energy levels available to absorb visible light so all the incoming light also exits and water appears as...what color?

Let's use fruit drink in a glass in your kitchen as an example to help us understand light absorption. The color of this particular fruit drink under the white light of your kitchen's light fixture is bright red; therefore, only red light is *exiting* the fruit drink glass, and it is this red light that your eyes see. Since the white light entering the fruit drink (from the overhead kitchen light fixture) contains all of the visible light's colors, yet the drink appears red, the molecules in the fruit drink must be absorbing ALL OF THE COLORS *EXCEPT* RED. Likewise, green limeade appears green when exposed to white light because the (usually artificial) dye in the drink absorbs all of the colors except green. And what color is green fruit drink when no light is shining on it? Right!

These light absorption properties can be used to determine the identity and concentration of matter too. Answer this question: Which red fruit drink has been over-diluted by my young son: the slightly pink glassful of liquid or the bright red glassful? Some red color tells you that both probably contain some fruit drink, but the weakly colored pink glass is not "red enough" and was therefore over-diluted by my son when he made it (he's only four!). Once again, the color tells you what was absorbed (everything but red) and the intensity tells you how much of the absorber is present (not much dye is present in the weak glassful and lots is present in the red glassful). In the case of the weak fruit drink, lots of the other light colors still pass through (are transmitted) and that makes the exiting color look only slightly red (pink). Remember that your eye's measurement of what you see basically also gives you a measure of what is missing too; lots of red light means the rest of the visible spectrum is missing.

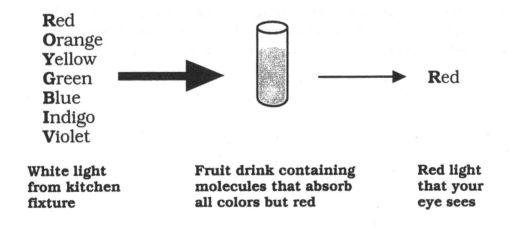

Red
Orange
Yellow
Green
Blue
Indigo
Violet

Red

**White light
from kitchen
fixture**

**Fruit drink containing
molecules that absorb
all colors but red**

**Red light
that your
eye sees**

Spectrophotometry

The lion's share of spectrophotometric chemical analysis is accomplished by basically this same technique. A light detecting instrument looks for a missing color and measures the amount that is missing compared to a standard. The more of a particular color that is absorbed, the more of the absorber must be present.

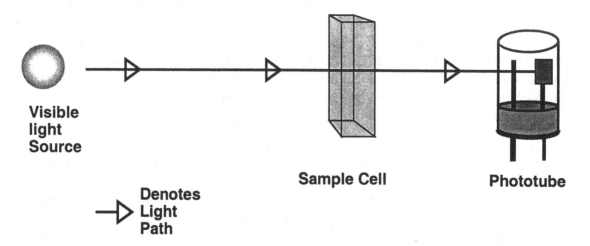

Visible light Source

Sample Cell

Phototube

Denotes Light Path

The figure above is a simplified diagram of a spectrophotometer. It contains all of the parts in the fruit drink experiment: a light source (a light fixture), a sample cell or cuvette (analogous to the glass of fruit juice) and a light detector (an electric eye called a phototube). You will probably use an instrument similar to this in the chromatography, lead, or fluoridation experiments in this laboratory manual. In Experiment 6: *Chromatography of Natural Pigments*, one of the final steps is to determine the amount of light absorbed at many different wavelengths (read colors) one by one. These data combined together and plotted on a graph make an *absorption spectrum*. Since the absorbances (or transmissions) in a spectrum for a given molecule depend on exactly what is absorbing the light and how much of it is there, spectrophotometry can be used to determine the presence and quantity of many components in the laboratory and in the atmosphere. A picture of a widely available spectrophotometer is in Experiment 18: *Fluoridation.*

Experiment 6

Chromatography of
Natural Pigments

Samples From Home

Bring about 50 grams (2 ounces) of recently unfrozen or fresh spinach that has been torn into
small pieces and sealed in a plastic bag.

Objectives

This experiment will introduce you to column chromatography and the separation of a few of the
natural pigments contained in spinach.

Background

The dark green color of spinach is actually a combination of many naturally colored substances.
This includes ß-carotene, which is one of many ingredients in spinach beneficial to humans. This
chemical is considered a vitamin A precursor because it is converted into vitamin A in our bodies.
ß-carotene is so successful as a natural color additive that the American food industry has adopted
it as a dye for producing various shades of red and yellow in foods. This in turn is beneficial to
consumers because it eliminates the need for a truly artificial color. The organic molecular "stick
structure" of ß-carotene is shown below. The alternating double bonds in the long chain that
connects the two rings helps to give the molecule its color. See Figure 6.1.

When isolated from a solution of petroleum ether, ß-carotene's crystals are red; however, when still dissolved in dilute solutions of this solvent, it exhibits a yellow color. Beta carotene is one of the natural pigments that you will isolate from spinach in this experiment. You will examine the color of this substance (while it is dissolved in a solvent) using your eye and an instrument called a spectrophotometer. Read *Introduction to Spectroscopy* located between Labs 5 and 6 in this laboratory manual if you have not done so already.

Other colored components of spinach include a group of molecules called chlorophylls, and the second component that will be separated in this experiment is a mixture of these compounds. Chlorophylls are organic molecules that coordinate (hold on to) magnesium in the center of a relatively complex chemical ring system. This is an example of an organometallic compound: a metal (magnesium) along with the normal elements in all organic molecules, carbon and hydrogen.

Figure 6.1 ß-carotene molecular structure.

Besides the important part chlorophylls play in the photosynthetic cycle of green plants, they are also extracted from plants and used to dye leather and as deodorants; however, their deodorant ability apparently has little effect on spinach.

Procedure

A. *Chromatographic separation of ß-carotene and chlorophylls*

CAUTION: *Petroleum ether and acetone are very flammable liquids. Use caution. Silica should not be breathed.*

1. Place your room temperature (freshly thawed is best) spinach in about 50 mL of a 80:20 mixture of petroleum ether/acetone in a mortar. Grind with the pestle until the liquid is dark green. Decant the solvent (separate the liquid from the solid by slowly pouring the liquid into a beaker) and place 5 mL of this extract in a test tube and centrifuge for 2 minutes. (Check with the lab instructor as to the size test tube that fits your centrifuge. Measure the volume of the 5 mL using a small graduated cylinder. If you have not been instructed in the use of the centrifuge then ask the lab instructor to demonstrate. Remember to use a counterweight tube for proper balance in the centrifuge. Set the centrifuged liquid aside for Step 6.

Put 50 mL of petroleum ether and 50 mL of acetone from the stock containers into two separate 100 mL beakers. You can use the markings on the beakers to estimate the volumes.

2. Secure a disposable Pasteur pipet in a buret clamp attached to a ring stand, taking care not to clamp and break the fragile tip of the capillary. Use a rubber band if necessary. If you break the pipet then put the pieces in the appropriate container and begin again.

3. Soak a swab of glass wool (balled up to be about the diameter of a dime) in petroleum ether and then, with a piece of wire, push the wool through the top of pipet down to the beginning of the tip of the Pasteur pipet, just where the body of the pipet starts to narrow. See Figure 6.2. Put enough sand in the pipet secured to the ring stand to make a layer about 0.25 cm over the glass wool.

4. Add silica to this chromatographic column with a spatula until you have a layer about 5 cm high. This can best be accomplished by attaching a 1 inch piece of rubber tubing and a plastic funnel to the top of the pipet. Add the silica slowly into the funnel and tap it gently to help prevent clogging. *Take precaution when handling silica. Do not breathe this solid!* Above the 5 cm (2 inches) column of silica, place another 0.25 cm ($^1/_8$ inch) layer of sand, again using the funnel set-up.

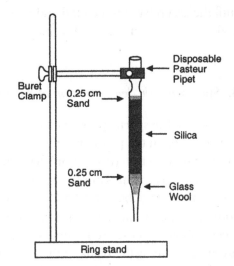

Figure 6.2. Chromatographic column.

Read the following and Steps 5 and 6 before you go on.

Fill the column with petroleum ether using a Pasteur pipet. At this point the column should drip at about 1 drop per second. If the dripping rate is unsatisfactory (3 drops per second is probably too fast and 1 drop every 5 seconds is too slow), stop, empty the solid material into an appropriate waste beaker and construct another column starting at Step 2.

5. Allow the chromatographic column to drip petroleum ether until the solvent level drops down to the top of the upper sand layer but no farther. Immediately go to Step 6. Collect all of the waste in this experiment in a 250 or 150 mL beaker and, at the end of the lab, empty it into a suitable waste crock or class waste container.

6. Quickly add enough spinach extract to fill the column to the top of the pipet.

7. Allow the column to drip until the spinach extract level falls to the top of the upper layer of sand. At this point fill the column again with **petroleum ether** (not more spinach extract) and keep the level of this solvent above the upper sand level throughout the steps described below. *Watch carefully so that the solvent level never drops below the topmost sand layer.*

8. Observe the chromatographic column as your chromatogram develops. Two colored bands will separate from the original green spinach mixture. As the lower, yellow band moves downward and then exits the tip of the column, collect this fraction in a 100 mm (4 inch) test tube. (You can use a small beaker if you wish. Clearly label whatever you use with wax pencil.) This yellow band is ß-carotene. When the petroleum ether level drops down to the upper sand layer and **after the yellow fraction has been collected,** fill the column to the top with **acetone** instead of

petroleum ether as before. Continue replacing the solvent at the top of the column with acetone until the green band has eluted and been collected in another similarly labelled test tube or small beaker. The green band is the chlorophyll fraction.

B. *Spectrometric determination of ß-carotene and chlorophylls*

1. Turn on the spectrophotometer and let it warm up. Put about 4 mL of the yellow fraction that you collected in the procedure above in one of the test tube-like cuvettes that come with the spectrophotometer. (If you don't have enough, pool multiple fractions from different students.) This is the *ß-carotene sample.* Always wipe your newly filled cuvettes with a Kim-Wipe or soft paper towel. Put about 4 mL of petroleum ether in another cuvette. This is the *blank. With nothing* in the sample compartment (also called the observations cell) adjust the spectrophotometer's zero knob until the meter reads exactly zero % transmission. Place the cuvette containing the blank in the sample compartment of the spectrophotometer. Make sure that the line on the cuvette always faces the same direction when you put the cuvette in the cell at different steps of the procedure. Close the cover.

2. Set the wavelength dial to 700 nanometers, nm (or 700 mμ, millimicrons). Adjust the transmittance/absorbance dial so that the meter shows 100% transmittance.

3. Remove the blank cuvette and insert the sample tube into the observation cell. Again make sure that the cuvette's line faces the correct direction. Close the cover. Read the absorbance off the meter with your sample in the cell. Get your lab instructor to help you estimate the last digit correctly. Record this absorbance on the report sheet adjacent to the correct wavelength settings.

4. Remove the sample cuvette and replace it with the blank cuvette. Adjust the wavelength dial to read 690 nanometers and adjust the transmittance dial to read 100% transmittance on the meter. Remove the blank cuvette, and again replace it with the sample cuvette, taking care to adjust the position of the cuvette's line as before. Read the sample absorbance from the meter and record this value.

5. Repeat this process for successively smaller wavelengths, decreasing (decrementing) the wavelength by 10 nanometers each time and ending at 400 nanometers. Don't mix up the sample and blank cuvettes or readings. If one person exchanges the samples and another person records the values, fewer mistakes will occur. Readings taken by a team can be shared on two different report sheets.

6. Repeat this procedure for the green chlorophyll fraction, **except put acetone in the blank cuvette instead of petroleum ether.** Start at 700 nm and decrement 10 nm a time down to 400 nm. Record the data on the data sheet in the correct column. Again, don't confuse the sample and blank cuvette's or readings.

7. Graph the data for both fractions on the two different pieces of graph paper provided in this lab. Each has been appropriately titled and the X-axis has already been numbered from 400 to 700 nanometers. Number the Y-axis with numbers starting just below the lowest absorbance reading that you took and ending just above your highest absorbance reading divided evenly. After

numbering your graph, plot the data by putting a dot on the graph at the intersection of the wavelength and absorbance readings for each data point you collected. Repeat this procedure for the chlorophylls data.

For the ß-carotene graph join the dots together by a straight line between each dot. This will yield a rough example of an absorption spectrum that could be obtained with a scanning spectrophotometer (an instrument that does the wavelength decrementing automatically and plots the result smoothly). The shape of this spectrum is characteristic of ß-carotene in petroleum ether solution and would be similar no matter which instrument is used to take the spectrum. This fingerprint can be useful for identification purposes. A spectrum of a molecule not examined in this experiment can be seen in the figure below.

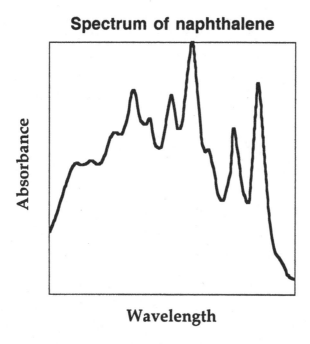

Figure 6.2 Spectrum of a molecule.

Chromatography of Natural Colors

Date _____   **Section number** _____   **Name** _____

Record the absorption readings on the same line as the wavelength.

Wavelength (in nanometers)	Absorption for ß-carotene	Absorption for chlorophyll
700	_____	_____
690	_____	_____
680	_____	_____
670	_____	_____
660	_____	_____
650	_____	_____
640	_____	_____
630	_____	_____
620	_____	_____
610	_____	_____
600	_____	_____
590	_____	_____
580	_____	_____
570	_____	_____
560	_____	_____
550	_____	_____
540	_____	_____
530	_____	_____
520	_____	_____
510	_____	_____
500	_____	_____
490	_____	_____
480	_____	_____
470	_____	_____
460	_____	_____
450	_____	_____
440	_____	_____
430	_____	_____
420	_____	_____
410	_____	_____
400	_____	_____

Chromatography of Natural Colors

Date _____   **Section number** _____   **Name** _____

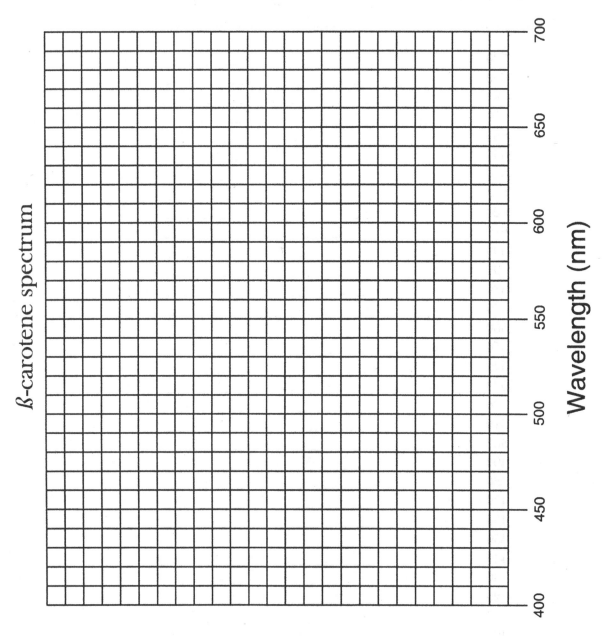

ß-carotene spectrum

Absorbance

Wavelength (nm)

Chromatography of Natural Colors

Date _____   Section number _____   Name _____

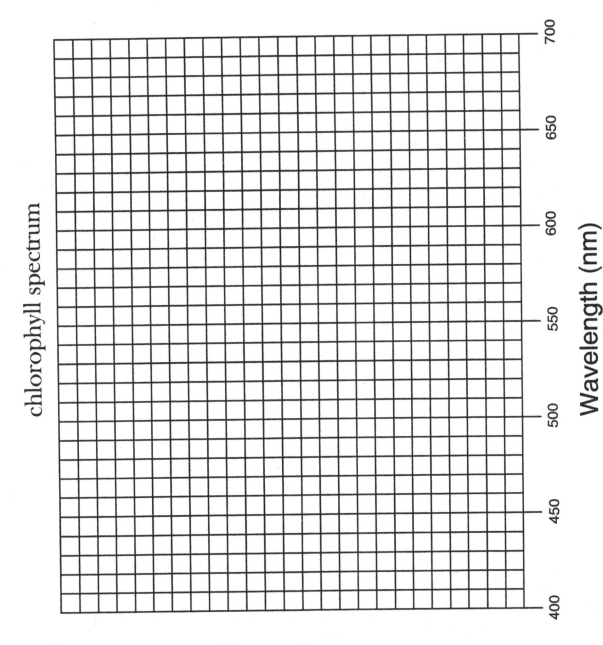

chlorophyll spectrum

Wavelength (nm)

Absorbance

Chromatography of Natural Colors

Date _____   **Section number** _____   **Name** _____

1. Why do many people consider that naturally derived dyes or pigments are safer for human beings than synthetically derived dyes when used to color food?

2. What would happen if only naturally derived dyes were allowed to be used in food sold in the United States? How would this affect the prices of foods with dyes in them?

3. Why is the green band referred to as chlorophylls instead of just chlorophyll?

Think, Speculate, Reflect, and Ponder

4. What would happen if acetone were the first solvent used to *elute* the dyes instead of petroleum ether? *Elution* is the process of separating or exiting the column one after the other.

5. Can you tell the color of ß-carotene by looking at your plot of absorbance versus wavelength for your ß-carotene fraction? How?

6. Can you relate the relative polarity of acetone and water to the structures found in a reference like *The Merck Index* or *CRC Handbook of Chemistry and Physics*? How?

Experiment 7

Heat of Combustion

Sample From Home

No sample from home is required for this experiment.

Objectives

In this experiment you will determine the heat of combustion of three fuels through the experimental technique called calorimetry. Based on your results, you will calculate which of these fuels is the most energy efficient.

Background

In the burning of hydrocarbons (combustion), carbon combines with gaseous oxygen (O_2) to form carbon dioxide, and hydrogen combines with O_2 to form water. The overall reaction releases chemical energy as heat. This heat is used to generate electricity, heat buildings, power automobiles or Bunsen burners, and a host of other jobs.

In 2000, coal accounted for approximately 51% of the fuel used in the United States; methane (natural gas) and nuclear power plants play a much smaller role as energy sources (13.5 and ~20% respectively). Cleaner energy producers such as hydrothermal, solar, and wind powered generators also contributed in a small but growing way to how Americans supply their increasing thirst for energy. In the Third World, a much larger fraction of the fuel burned is wood or charcoal (charred wood) because these are the only sources available both physically and economically—that is, what most third world peoples can afford. The ready availability of these sources is shrinking, however, as the population of the underdeveloped nations continues to increase. One of the most difficult jobs in this new century will be supplying energy to the growing populations of the world in such a way that the polluting mistakes of the past are not repeated.

An important property of any fuel is its heat of combustion or the amount of energy released for every gram or mole of fuel burned. (A mole is a chemical unit defining a certain number of molecules.) The heat of combustion determines, in part, the value of a fuel to the user. Other important factors include how cleanly the fuel burns and how convenient the fuel is to store, transport, and introduce into the combustion chamber of the device in which it will be burned. These aspects of fuels—heat of combustion, burning characteristics, and "storeability"—make it very important to examine the range of fuels that are available for use on our planet and to choose those that have other desirable characteristics besides merely the availability and the present market cost.

Calorimetry

Calorimetry is the measurement of the heat released (or absorbed) in a chemical reaction. Heat generated by the burning of a fuel is captured by an instrument called a calorimeter and the amount of this energy is measured. The amount of energy generated can be determined by measuring how much the temperature of a container of water changes when a known amount of fuel is burned to heat it. With *your* calorimeter, you will determine the amount of heat released from burning a particular fuel by measuring the temperature increase of a known mass of water when it is heated by burning a known amount of fuel. The amount of temperature increase will depend on the specific heat of water, the specific heat of the flask containing it, and the heat of combustion of the fuel. The specific heat of a substance is the amount of heat it takes to raise the temperature of one gram of that substance by one degree Celsius and is a measure of a substance's ability to absorb and store heat. It is a characteristic physical property just like melting point or density.

For example, the specific heat of water is 1 calorie per gram per degree Celsius (1 cal/g-° C): It takes 1 calorie of energy to raise the temperature of one gram of water by one degree Celsius. More water (more grams) or more temperature increase (a greater rise in temperature of a sample) would require more energy (more calories). The specific heat of the Pyrex™ glass making up the Erlenmeyer flask holding the water is much less than that of water: 0.205 cal/g-° C. Therefore, it takes less energy to raise the temperature of the flask 1 degree than it does to raise the temperature of the same mass of water by one degree. When you perform this experiment, the energy released by burning the fuel will heat both the water and the flask containing that water (we assume by the same amount) so both the specific heat of water and the specific heat of the flask must be taken into account.

Procedure

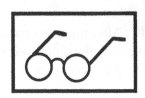

A. Heat of combustion of methanol

1. Weigh a clean and dry 250 mL Erlenmeyer flask to 0.01 g and record this mass on the report sheet.

2. With the flask still on the balance pan, add tap water to the flask until the balance reads about 200 g of *added* mass. Record the exact mass of the flask plus the water on your report sheet in the appropriate blank. By subtracting the datum from Step 1 from that in Step 2 you can determine the exact mass of water that you added to the flask. Record this mass, obtained by subtraction, in the correct place on the report sheet.

3. Prepunched coffee cans will be available for your use as a calorimeter. Set up your calorimeter by carefully inserting the flask containing the water through the hole in the coffee can bottom and securing the flask to the ring stand with a clamp (refer to Figure 7.1). Be careful of the sharp edges of the cut out metal bottom! The can will completely cover the lower part of the flask with about an inch to spare.

4. Put a small piece of rubber tubing over the tip of your thermometer's bulb to protect it during stirring.

5. Place 20 to 30 mL of methanol in your alcohol burner. Don't pull the wick up out of the wick holder any more than it already is! Your lab instructor has adjusted the wick already for the best flame height.

6. Place the cap on the alcohol burner and weigh to 0.01 g. Record this mass on your report sheet.

Figure 7.1. Calorimeter apparatus.

7. Record the starting temperature of the water in the Erlenmeyer to the nearest 0.1 ° C. The thermometer will now remain in the water in the flask for the remainder of the experiment.

8. Place the burner beneath the Erlenmeyer flask making sure that the coffee can will at least cover the flame during the experiment. Remember the idea is to catch *all the heat that you can from the burning flame*; however, if you lower the can/flask too much it will smother the flame.

9. Light the alcohol burner with a match.

10. As you heat, continually stir the water with the thermometer until a 20 ° C change in temperature has occurred as measured by the thermometer in the water. You can monitor the flame by watching its reflection in the neck of the flask. If the flame goes out, quickly relight it with another match.

11. After the temperature has risen by about 20 ° C, blow out the flame and continue to stir and measure the temperature until it stops rising. Record the maximum temperature. This may take 20 or 30 seconds after you blow out the flame.

12. Remove the burner from under the coffee can and reweigh it with its cap on. Record the weight to 0.01 g.

13. You now have recorded on your report sheet the mass of water that you heated, the temperature increase of that water, and the mass of fuel it took to cause this increase.

14. To get your duplicate set of data, repeat Steps 1 through 12.

B. Heat of combustion of candle wax

1. Exactly as before, weigh a dry Erlenmeyer flask (record this mass), add approximately 200 g water, and determine the weight of the water that you added to 0.01 g. Again, you will have to subtract the mass of the empty flask from the mass of the flask plus water. Reassemble the calorimeter as before being careful not to cut yourself on the sharp metal edges!

2. Get a candle from the stock room or supply cart. Light the candle and drip hot wax onto the center of the watch glass until the bottom of the candle can be stuck securely to the watch glass. Blow out the candle and allow it to cool and the wax to harden.

3. Weigh the watch glass and candle to 0.01 g. Record this mass. Record the starting temperature of the water to the nearest 0.1 °C.

4. Position the candle under the coffee can as before and light the candle. Make sure that the candle's flame is not snuffed out by the bottom of the flask.

5. Stirring continuously and exactly as before, heat the water in the flask until a 20 ° C temperature rise has occurred. Record the highest temperature achieved by the water *after* you blow out the candle.

6. Reweigh the candle and watch glass to 0.01 g and record this mass.

7. You now have recorded on your report sheet the mass of water heated, its increase in temperature, and the mass of candle that was needed to cause this rise. Repeat Steps 1 through 6 for a second run.

Figure 7.2. Candle setup.

C. Heat of combustion of wood

1. Repeat the apparatus assembly. Make sure that you get the weight of the clean and dry Erlenmeyer first. Then get the weight of this flask containing approximately 200 g of water exactly as before. Record these masses as well as the calculated mass of the added water.

2. Get 20 wooden splints from the stock room or supply cart. Put the splints on top of a wooden block and weigh the splints *and* the block to 0.01 g. Carefully arrange the splints in a 5 or 6 level campfire (or stacked like a log cabin) on top of the wooden block so that when lit, as many of the splints will burn as possible. Record the temperature of the water to 0.1° C.

3. Place the splint campfire underneath the coffee can. You may need to raise the flask/can assembly by moving the clamp up the ring stand. *Don't forget to retighten the clamp after you move it.* After the fire is lit, carefully reposition the calorim-

Figure 7.3 "Campfire" of wooden splints.

eter so that as much of the wooden block and the campfire is covered as possible.

4. Light the campfire in numerous places using a wooden match, allowing the fire to burn toward the middle. Relight the fire if necessary. You may have to adjust the coffee can's height to be able to get to the edges of the campfire with your match.

5. Constantly stirring the water with the thermometer, record the highest temperature that the water temperature rises to *after the wood fire burns out.* Note: If the fire burns too slowly or too quickly the experimental results will have a large error. You may have to experiment with the configuration of your campfire to get it to burn in such a way that lots of the heat doesn't escape the confines of your calorimeter. Watch how other students in your lab design their campfires and adopt the best configuration. Hint: If you see lots of soot on the *outside* of your coffee can, your campfire probably burned too quickly.

6. After the fire has burned out, reweigh the wooden block and (all the pieces of the burned splints) to 0.01 g. Take care not to spill or lose any of the burned residue. Put any pieces that fall off of the block back on the block before you reweigh.

7. You now have recorded on your report sheet the mass of water, the temperature increase, and the mass of wood necessary to cause this temperature rise. Repeat Steps 1 through 6.

Calculations

A. Heat of combustion of methanol

The amount of heat absorbed by the calorimeter (Erlenmeyer + water) when methanol was burned may be calculated from the masses of the Erlenmeyer and water, the temperature change, and the respective specific heats. Repeat the calculation for each trial separately and then average the heat of combustion for each trial at the end. Remember that for a particular trial, the temperature change for the flask will be the same as the temperature change for the water.

Heat absorbed by Erlenmeyer flask (in calories) = (mass of flask) X (0.205 cal/g-°C) X (temperature change):

Mass of empty flask   _____g trial 1.

                                                    _____g trial 2.

Temperature change   _____° C trial 1.

                                                      _____° C trial 2.

Specific heat of pyrex glass   0.205 cal/g-° C

Heat absorbed by Erlenmeyer flask   _____calories trial 1.

                                                      _____calories trial 2.

Heat absorbed by water (in calories) = (mass of water) X (1 cal/g-°C) X (temperature change):

Mass of water   _____g trial 1.

                                                    _____g trial 2.

Temperature change   _____° C trial 1.

                                                      _____° C trial 2.

Specific heat of water   1 cal/g-° C

Heat absorbed by water   _____calories trial 1.

                                                      _____calories trial 2.

Total heat absorbed by calorimeter = heat absorbed by Erlenmeyer flask + heat absorbed by water

_____ calories trial 1.

_____ calories trial 2.

Heat of combustion of methanol = heat absorbed by calorimeter divided by the mass of methanol burned

Mass of methanol burned

_____grams trial 1.

_____grams trial 2.

Heat of Combustion   _____calories/gram trial 1.

_____calories/gram trial 2.

Average heat of combustion for methanol   _____calories/gram.

(record on Report Sheet)

B. Heat of combustion of candle wax

The amount of heat absorbed by the calorimeter (Erlenmeyer + water) when candle wax was burned may be calculated from the masses of the Erlenmeyer and water, the temperature change, and the respective specific heats.

Heat absorbed by Erlenmeyer flask (in calories) = (mass of flask) X (0.205 cal/g-°C) X (temperature change):

Mass of flask   _____g trial 1.

_____g trial 2.

Temperature change   _____° C trial 1.

_____° C trial 2.

Specific heat of pyrex glass   0.205 cal/g-° C

Heat absorbed by Erlenmeyer flask   _____calories trial 1.

_____calories trial 2.

Heat absorbed by water (in calories) = (mass of water) X (1 cal/g-°C) X (temperature change):

Mass of water   _____g trial 1.

                                           _____g trial 2.

Temperature change   _____° C trial 1.

                                           _____° C trial 2.

Specific heat of water   1 cal/g-° C

Heat absorbed by water   _____calories trial 1.

                                           _____calories trial 2.

Total heat absorbed by calorimeter = heat absorbed by Erlenmeyer flask + heat absorbed by water

                                           _____calories trial 1.

                                           _____calories trial 2.

Heat of combustion of candle wax = heat absorbed by calorimeter divided by the mass of candle burned

                                           _____calories/gram trial 1.

                                           _____calories/gram trial 2.

Average heat of combustion of candle wax   _____calories/gram.
                                           (record on Report Sheet)

C. Heat of combustion of wood

The amount of heat absorbed by the calorimeter (Erlenmeyer + water) when wood was burned may be calculated from the masses of the Erlenmeyer and water, the temperature change, and the respective specific heats.

Heat absorbed by Erlenmeyer flask (in calories) = (mass of flask) X (0.205 cal/g-°C) X (temperature change):

Mass of flask   _____g trial 1.

                                           _____g trial 2.

Temperature change   _____° C trial 1.

                                           _____° C trial 2.

Specific heat of pyrex glass   0.205 cal/g-°C

Heat absorbed by flask   _____calories trial 1.

                                                               _____calories trial 2.

Heat absorbed by water (in calories) = (mass of water) X (1 cal/g-°C) X (temperature change):

Mass of water   _____g trial 1.

                                                               _____g trial 2.

Temperature change   _____° C trial 1.

                                                               _____° C trial 2.

Specific heat of water   1 cal/g-°C

Heat absorbed by water   _____calories trial 1.

                                                               _____calories trial 2.

Total heat absorbed by calorimeter = heat absorbed by flask + heat absorbed by water

                                                               _____ calories trial 1.

                                                               _____ calories trial 2.

Heat of combustion of wood = heat absorbed by calorimeter divided by the mass of wood burned

                                                               _____calories/gram trial 1.

                                                               _____calories/gram trial 2.

Average heat of combustion of wood   _____calories/gram.
                                                               (record on Report Sheet)

Heat of Combustion

Date _____   Section number _____   Name _____

Fuel ---->	Methanol	Methanol	Candle	Candle	Wood	Wood
Trial # ---->	# 1	# 2	# 1	# 2	# 1	# 2
Mass of Flask + water						
Mass of empty flask						
Mass of water						
Final water temperature						
Initial water temperature						
Temp. change						
Initial mass of burner + cap						
Final mass of burner + cap						
Mass of methanol burned						
Initial mass candle + watch glass						
Final mass candle + watch glass						
Mass of candle burned						
Initial mass wooden block + splints						
Final mass block + burned splint						
Mass of wood burned						

Average Heat of Combustion			
	cal/gram	cal/gram	cal/gram

Heat of Combustion

Date _____   **Section number** _____   **Name** _____

1. Which of the three fuels that you used yields the most energy per gram of fuel? List all three average values that you calculated for the three fuels.

2. Number the following substances in order of increasing specific heats: Aluminum, Argon, Beryllium, Copper, Helium, Hydrogen, Lead, and Xenon . You will need specific heat data for this problem from a reference like the *CRC Handbook of Chemistry and Physics.*

3. What other factors besides energy released per gram of fuel determine the usefulness of a fuel?

4. What might be the largest source of error in this experiment? Explain your answer.

Think, Speculate, Reflect, and Ponder

5. Are there any environmental factors that are taken into account when choosing a fuel? If so what are they and why *should* they be taken into account when choosing a fuel?

6. Why do you have to calculate the heats of combustion of wood and the candle on a per gram basis instead of a per mole basis?

7. Calculate the heat of combustion of methanol on a per mole basis assuming methanol has a molecular weight of 32.05 grams/mole. Use the data on your report sheet.

Experiment 8

Unknowns in an Homologous Series

Samples From Home

No sample from home is required for this experiment.

Objectives

Using calorimetry, this experiment will allow you to compare the different potential energies of a family of fuels that structurally vary only in carbon chain length: Each member of the family of alcohols that you will study is identical except for the number of CH_2 units it contains. The relationship between carbon number and heat of combustion will allow you to determine the identity of an unknown member of the series.

Background

In a sense, all of the energy sources that we use on earth are derived from solar energy. Even the radioactive elements that fuel nuclear power plants ultimately came from stars. Energy from the sun in the form of light is captured by plants by the formation of chemical bonds. Later these bonds are broken and their energy is released in the form of chemical or heat energy (this is called cellular respiration).

The process of combustion (or burning) is similar to respiration in that high energy, less stable molecules react and form lower energy, more stable products, thereby giving up energy that can be used for some purpose. Wood, containing stored energy, burns in the presence of oxygen, and the results are lower energy, more stable products plus heat and light given off. This is, of course, true for all fossil fuels such as methane, coal, and petroleum. In the case of most combustion, human beings are interested in capturing the energy derived from the burning of the fuel in order to do something useful such as powering an automobile or heating a home. In the case of cellular respiration, organisms are "interested" in powering the processes necessary for life: reproduction, growth, movement, etc. All fuels are not equivalent, however. Experiment 7: *Heat of Combustion* compared the energy contained in three fuels that were decidedly unequal in the amount of energy that each supplied per gram (differing heat contents). The differences in these three fuels are in great part due to their differing chemical structures.

Methane (CH_4), ethane (CH_3-CH_3), and propane (CH_3-CH_2-CH_3) are homologues (also spelled homologs). This means that they differ from each other only in the number of CH_2 groups they contain. (See Experiment 26: *Chemical Model Building*.) In this experiment, a series of homologues will be burned, the energy that is released captured, and the heats of combustion of each compound measured. You will be able to determine the structure of an unknown member of this group by comparing its heat of combustion to the heats of combustion of other, known members of the series.

The workhorse in this experiment will be your alcohol burner. The fuels that you will examine are all alcohols (they contain a single -OH group). You will use your burner to study four different alcohols over the course of the experiment, each time using a different wick that is specified for each fuel. It is very important that wicks saturated with one fuel not be used with other fuels. This will yield an unacceptable error in the heat of combustion that you determine because the wicks retain a relatively large amount of fuel from the last burner they were in.

To make sure that we avoid this problem, wicks that have been used in methanol (an alcohol containing 1 carbon atom) will be deposited into a beaker labelled METHANOL WICKS; wicks that have been used with unknown #1 will be deposited in a beaker labelled UNKNOWN #1 WICKS; propanol (an alcohol containing 3 carbon atoms) wicks deposited in a beaker labelled PROPANOL WICKS, etc. If you need a wick for a pentanol run (pentanol is an alcohol containing 5 carbon atoms), take out the wick that is in your burner and put it in the correctly labelled beaker and get a pentanol wick from the PENTANOL wick beaker.

When you are inserting a new wick into the wick holder of the alcohol burner, insert the new wick *up from the bottom* and do not pull the wick more that a few millimeters above the top of the wick holder. *If you do* extend the wick too much, your flame will be too tall and this will cause an error in your results because you will lose too much heat during the experiment—heat that you want to catch with your calorimeter (see Experiment 7: *Heat of Combustion*). Please ask your lab instructor if you have a question about the heights of your wicks.

Since you will be emptying and adding different alcohols to your burner, it is important that you empty the burner of the first fuel as completely as possible before you add the next fuel. Pour the leftover fuel from your alcohol burner into the bottle containing the fuel with the correct name: The bottle labelled PENTANOL FUEL will contain only leftover pentanol; the bottle labelled UNKNOWN ALCOHOL # 2 will contain only leftover unknown alcohol #2, etc. Be careful when you empty and refill your burner so the beakers will only have the correct fuels in them.

Similarly, when it is time to put a new fuel in your burner make sure that you "fuel-up" from the bottle containing the correct fuel. If you have any question about which wick beaker or fuel bottle is which, ask your lab instructor. Finally, it is not necessary to completely fill your burner with fuel. Only 20 to 30 mL of fuel is necessary for two trials with each fuel.

Procedure

The procedures and calculations for this experiment are identical to those of the last calorimeter experiment, Experiment 7: *Heat of Combustion*. Reviewing the introduction to that experiment will help you with this one.

A. Heat of Combustion of Methanol

1. Weigh a clean and dry 250 mL Erlenmeyer flask to 0.01 g and record this mass on the report sheet.

2. With the flask still on the balance pan, add tap water to the flask until the balance reads about 200 g of *added* mass. Record the exact mass of the flask plus the water on your report sheet in the appropriate blank. By subtracting the datum from Step 1 from that in Step 2 you can determine the exact mass of water that you added to the flask. Record this mass, obtained by subtraction, in the correct place on the report sheet. If you use a balance that tares (ask your lab instructor if you don't know), record the mass of fuel you added on the report sheet.

3. Set up your coffee can calorimeter by carefully inserting the flask containing the water through the hole in the coffee can bottom and securing the flask to the ring stand with a clamp (refer to Figure 8.1). Be careful of the sharp edges of the cut-out metal bottom! The can will completely cover the lower part of the flask with about an inch to spare.

4. Put a small piece of rubber tubing over the tip of your thermometer's bulb to protect it during stirring.

5. Place 20 to 30 mL of *methanol* in your alcohol burner. Don't pull the wick up out of the wick holder any more than it already is! Your lab instructor has adjusted the wick already for the best flame height.

6. Place the cap on the alcohol burner and weigh to 0.01 g. Record this mass on your report sheet.

7. Record the starting temperature of the water in the Erlenmeyer to the nearest 0.1 °C. The thermometer will now remain in the water in the flask for the rest of the experiment.

8. Place the burner beneath the Erlenmeyer flask making sure that the coffee can will at least cover the flame during the experiment. Remember the idea is to catch *all the heat that you can from the burning flame*; however, if you lower the can/flask too much it will smother the flame.

9. Light the alcohol burner with a match. As you heat, continually stir the water with the thermometer until approximately 20 ° C change in temperature has occurred as measured by the thermometer in the water. You can monitor the flame by watching its reflection in the neck of the flask. If the flame goes out, quickly relight it with another match.

10. After the temperature has risen by 20 ° C, blow out the flame and continue to stir and measure the temperature until it stops rising. Record the maximum temperature. This may take 20 or 30 seconds after you blow out the flame.

11. Remove the burner from under the coffee can and reweigh it with its cap on. Record the weight to 0.01 g.

12. Repeat Steps 1 through 11 without adding any more fuel. You will need to start Step 1 with a dry, *room temperature* flask into which you add tap water.

Figure 8.1. Calorimeter apparatus.

B. Heat of Combustion of Propanol

Repeat Procedure A *with propanol instead of methanol as the fuel.* Make sure that when you are emptying and replacing the fuel in your burner you use the correct wick beaker and fuel bottle: Place the old wick in the beaker labelled METHANOL WICKS and pour any leftover methanol in the bottle labelled METHANOL FUEL. Refill your burner with propanol from the bottle labelled PROPANOL FUEL and position a wick from the beaker labelled PROPANOL WICKS in your wick holder. Perform two trials with this fuel and record your data on the report sheet.

C. Heat of Combustion of Pentanol

Repeat Procedure A with pentanol. Make sure when you are emptying and replacing the fuel in your burner that you use the correct wick beaker and fuel bottle. Also, do the experiment in duplicate (two trials) as before.

D. Heat of Combustion of an Unknown

Choose one or the other of the available unknowns and repeat the steps in procedure A (in duplicate) with the unknown alcohol as your fuel. (Your lab instructor may prefer to assign unknowns: ask.)

Don't forget to record your unknown number on the report sheet. When you finish, you will have two trials for each of the known alcohols and one of the unknown alcohols.

Calculations

A. Calculating the Heats of Combustion

The amount of heat absorbed by the calorimeter (Erlenmeyer + water) when each fuel was burned may be calculated from the masses of the Erlenmeyer and water, the temperature change, and the respective specific heats. The specific heat of a material is characteristic of that material and is the amount of heat necessary to raise the temperature of 1 gram of the material by 1 degree (cal/g-°C) (see Experiment 7: *Heat of Combustion*). Repeat the calculation for each fuel for each trial separately and then average the heat of combustion for each trial at the end. Remember that for a particular trial, the temperature change for the flask will be the same as the temperature change for the water.

Heat of combustion for methanol

Heat absorbed by Erlenmeyer flask (in calories)
(multiply the mass of the Erlenmeyer times 0.205 cal/g-°C then multiply the result by the temperature change):

Mass of empty flask   _____g trial 1.

_____g trial 2.

Temperature change   _____° C trial 1.

_____° C trial 2.

Specific heat of Pyrex™ glass   0.205 cal/g-° C

Heat absorbed by Erlenmeyer flask   _____calories trial 1.

_____calories trial 2.

Heat absorbed by water (in calories)
(multiply the mass of water times 1.00 cal/g-°C them multiply the result by the temperature change):

Mass of water

        _____g trial 1.

        _____g trial 2.

Temperature change

        _____° C trial 1.

        _____° C trial 2.

Specific heat of water

        1.00 cal/g-° C

Heat absorbed by water

        _____calories trial 1.

        _____calories trial 2.

Total heat absorbed by calorimeter
(add heat absorbed by Erlenmeyer flask and the heat absorbed by water)

        _____ calories trial 1.

        _____ calories trial 2.

Heat of combustion of methanol
(divide the heat absorbed by calorimeter by the mass of methanol burned)

Mass of methanol burned

        _____grams trial 1.

        _____grams trial 2.

Heat of combustion

        _____calories/g trial 1.

        _____calories/g trial 2.

Average heat of combustion for methanol

        _____calories/gram.

        (record on Report Sheet)

Heat of combustion of propanol

Heat absorbed by Erlenmeyer flask (in calories)
(multiply the mass of the Erlenmeyer times 0.205 cal/g-°C then multiply the result by the temperature change):

Mass of empty flask   _____ g trial 1.

                                                             _____ g trial 2.

Temperature change   _____ ° C trial 1.

                                                             _____ ° C trial 2.

Specific heat of Pyrex glass   0.205 cal/g-° C

Heat absorbed by Erlenmeyer flask   _____ calories trial 1.

                                                             _____ calories trial 2.

Heat absorbed by water (in calories)
(multiply the mass of water times 1.00 cal/g-°C then multiply the result by the temperature change):

Mass of water   _____ g trial 1.

                                                             _____ g trial 2.

Temperature change   _____ ° C trial 1.

                                                             _____ ° C trial 2.

Specific heat of water   1.00 cal/g-° C

Heat absorbed by water   _____ calories trial 1.

                                                             _____ calories trial 2.

Total heat absorbed by calorimeter
(add the heat absorbed by Erlenmeyer flask and the heat absorbed by water

                                                             _____ calories trial 1.

                                                             _____ calories trial 2.

Heat of combustion of propanol
(divide the heat absorbed by calorimeter by the mass of propanol burned)

Mass of propanol burned

                                                             _____ grams trial 1.

                                                             _____ grams trial 2.

Heat of combustion   _____calories/g trial 1.

_____calories/g trial 2.

Average heat of combustion for propanol   _____calories/gram.

(record on Report Sheet)

Heat of combustion of pentanol

Heat absorbed by Erlenmeyer flask (in calories)
(multiply the mass of the Erlenmeyer times 0.205 cal/g-°C then multiply the result by the temperature change):

Mass of empty flask   _____g trial 1.

_____g trial 2.

Temperature change   _____° C trial 1.

_____° C trial 2.

Specific heat of Pyrex glass   0.205 cal/g-° C

Heat absorbed by Erlenmeyer flask   _____calories trial 1.

_____calories trial 2.

Heat absorbed by water (in calories)
(multiply the mass of the water times 1.00 cal/g-°C) then multiply the result by the temperature change):

Mass of water   _____g trial 1.

_____g trial 2.

Temperature change   _____° C trial 1.

_____° C trial 2.

Specific heat of water   1.00 cal/g-° C

Heat absorbed by water   _____calories trial 1.

_____calories trial 2.

Total heat absorbed by calorimeter
(add the heat absorbed by Erlenmeyer flask and the heat absorbed by water)

_____ calories trial 1.

_____ calories trial 2.

Heat of combustion of pentanol
(divide the heat absorbed by calorimeter by the mass of pentanol burned)

Mass of pentanol burned

_____grams trial 1.

_____grams trial 2.

Heat of combustion

_____calories/g trial 1.

_____calories/g trial 2.

Average heat of combustion for pentanol

_____calories/gram.

(record on Report Sheet)

Heat of combustion of an unknown alcohol

Unknown #   _____

Heat absorbed by Erlenmeyer flask (in calories)
(multiply the mass of the Erlenmeyer times 0.205 cal/g-°C then multiply the result by the temperature change):

Mass of empty flask

_____g trial 1.

_____g trial 2.

Temperature change

_____° C trial 1.

_____° C trial 2.

Specific heat of Pyrex glass

0.205 cal/g-° C

Heat absorbed by Erlenmeyer flask

_____calories trial 1.

_____calories trial 2.

Heat absorbed by water (in calories)
(multiply the mass of the water times 1.00 cal/g-°C then multiply the result by the temperature change):

Mass of water   _____g trial 1.

                                                       _____g trial 2.

Temperature change   _____° C trial 1.

                                                       _____° C trial 2.

Specific heat of water   1.00 cal/g-° C

Heat absorbed by water   _____calories trial 1.

                                                       _____calories trial 2.

Total heat absorbed by calorimeter

(add the heat absorbed by the Erlenmeyer flask and the heat absorbed by water)

                                                       _____ calories trial 1.

                                                       _____ calories trial 2.

Heat of combustion of unknown
(divide the heat absorbed by calorimeter by the mass of unknown alcohol burned)

Mass of unknown alcohol burned

                                                       _____grams trial 1.

                                                       _____grams trial 2.

Heat of combustion   _____calories/g trial 1.

                                                       _____calories/g trial 2.

Average heat of combustion for unknown alcohol   _____calories/gram.

                                                       (record on Report Sheet)

B. Calibration Plot: Heat of Combustion per Gram Versus Carbon Number

On the graph paper provided, make a calibration plot of the heat of combustion of each of the known alcohols (in calories/gram) versus the number of carbons that each alcohol molecule contains (see below). Put the number of carbons on the x-axis and the heat of combustion per gram on the y-axis. For example, number your x-axis from 0 to 6 (carbons) and number your y-axis from 0 to 10,000 (calories per gram). Your experimental data points will fit on this graph as described.

Experimen

Methanol's structure is CH_3OH; it has one carbon per molecule. Propanol's structure is $CH_3CH_2CH_2OH$; it has 3 carbons. Pentanol's structure is $CH_3CH_2CH_2CH_2CH_2OH$, and it has 5 carbons per molecule. Your calibration plot of heat of combustion per gram versus the number of carbons in a molecule of the fuel will have three points: one for methanol, one for propanol, and one for pentanol.

Draw a straight line that visually best approximates the lie of these three points. If done well this line will be equidistant from all three of your points. *The best fit line will not pass through the intersection of the x-and y-axes.* You can place a long pencil or a ruler down on the graph and move it around until it passes an equal distance from all three. Finally draw a line on the graph that fits this line.

C. Determination of Unknown's Carbon Number

Plot the heat of combustion per gram of your unknown alcohol on the calibration plot. Determine the number of carbons in your unknown alcohol by relating the number of carbons to the heat of combustion per gram that you calculated. Here's how:

The calculated heat of combustion of your unknown will fall somewhere on the y-axis. Draw a *horizontal* line from the point on the y-axis that represents the calculated heat of combustion per gram of your unknown over to the best-fit line that you drew in the step above. Next draw a *vertical* line from that point on the best fit line down to the x-axis. This point on the x-axis should be close to the whole number of carbon atoms that your unknown alcohol contains. If this point is not exactly on a whole number then decide which whole number it lies *nearest* to on the x-axis. This is the number of carbons in your unknown alcohol. Remember: None of the number of carbons in the unknowns in this experiment is the same as any of the known alcohols. Write down the number of carbons in your unknown on the report sheet. If you have difficulty with this step, ask a classmate or your lab instructor for help.

Homologous Series

Date _____ Section number _____ Name _____

Fuel ------>	Methanol	Methanol	Propanol	Propanol	Pentanol	Pentanol	Your Unknown	Your Unknown
Trial #	# 1	# 2	# 1	# 2	# 1	# 2	Trial # 1	Trial # 2
Mass of Flask + water								
Mass of empty flask								
Mass of water								
Final water temperature								
Initial water temperature								
Temperature change								
Heat of Combustion								
Average Heat of Combustion	cal/gram		cal/gram		cal/gram		cal/gram	

Unknown number _____

Number of carbons in your unknown alcohol determined from the calibration plot ___ .

Homologous Series

Date _____ **Section number** _____ **Name** _____

Calibration Plot:
Number of carbons versus Heat of Combustion

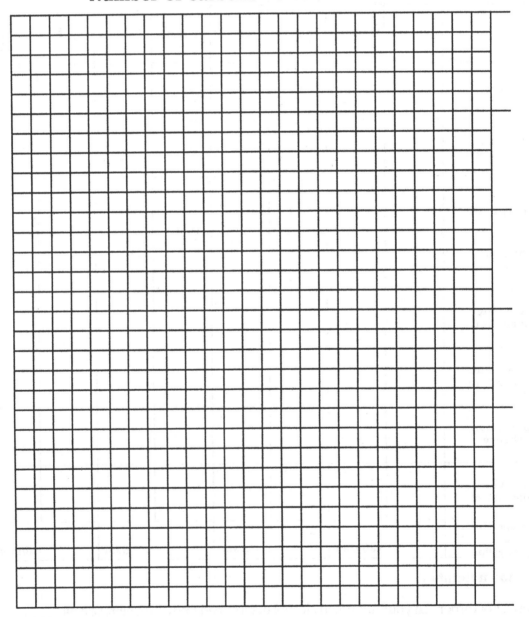

Number of Carbon Atoms per Molecule

Heat of Combustion (cal/gram)

Experim

Homologous Series

Date _____   **Section number** _____   **Name** _____

1. Could you see any visible difference between the flames of the different known alcohols? In other words, did carbon number seem to affect the colors of your flames? How?

2. If you answered yes to question 1, what would you expect the color of a flame that is burning a 10 carbon alcohol would be? If you answered no to question one, then reread the background to Experiment 1: *The Ubiquitous Bunsen Burner* and answer the following question: What effect does the ratio of the reactants mixing in the Bunsen burner's flame have upon the products of the reaction taking place?

3. What were the labels on the two axes of your graph?

Think, Speculate, Reflect, and Ponder

4. Why is the best fit line on your graph not drawn through all the points and/or why would the three experimental points that you derived from your data not always fall in a straight line?

5. Is there a way to calculate an equation that mathematically represents your best-fit line? How could you do this?

Experiment 9

Alcohol Content of Beverages & Consumer Products

Samples From Home

Bring 50 mL of a liquid containing water and ethyl alcohol as the only significant volatile components; examples might be after shave lotion, whiskey, cologne, wine, mouthwash, beer, etc. Check the label, or ask your lab instructor if in doubt. (NOTE: Cough syrups will not work, and rubbing alcohol may not be the same as ethyl alcohol.)

Objectives

The nature of distillation and the usefulness of a distillation apparatus for separating substances will be explored. Typical distillation behavior of two liquids will also be illustrated. An hydrometer or Westphal balance will be employed for determining the density of the distilled liquid, with the results used to verify or discover the actual alcohol content of the test sample.

Background

When relatively large amounts of liquids must be separated, purified or analyzed, the first technique a chemist thinks of is *distillation*. Moonshiners know about it, and so do the "reveenoorers." All liquids which have a reasonable tendency to evaporate (*volatile* liquids chemists would say) can be made to boil (to *distill*) if heated sufficiently under an appropriate pressure. If such a *volatile* liquid, like water, has mixed with it something else that is *nonvolatile* like salt, *distillation* will be able to separate the two components easily.

For example, boiling salt water will result in the volatile water coming off as steam, leaving the solid nonvolatile salt residue behind. In practice, the hot liquid vapors (steam in this case) are passed through a condenser in order to cool them sufficiently so that the steam will condense back into water. No chemical change has taken place—the salt is still salt, and the water is still water. If we were to pour the water back onto the salt residue, the original salt water mixture would form. We have simply performed a physical process or change.

The condenser itself simply consists of two concentric tubes—the inner one for your hot vapors to condense in, and the outer one in which to circulate cooling water. There is no opening joining the inner and outer tubes, and thus no way water can get inside of the inner tube. The liquid you will see dripping out of the end of your condenser will just be due to the hot distillate vapors condensing back into a liquid inside the inner tube.

But what about using distillation to separate two liquids which have different boiling points and thus different tendencies to evaporate? Even the oxygen and nitrogen used in laboratories and in commerce is isolated from the distillation of air—after it has first been turned into a liquid mixture containing these two elements by cooling to –196 °C. For our purposes, we need stress only this: although such separations can often (but not always) be successfully performed, they are more difficult than it might seem.

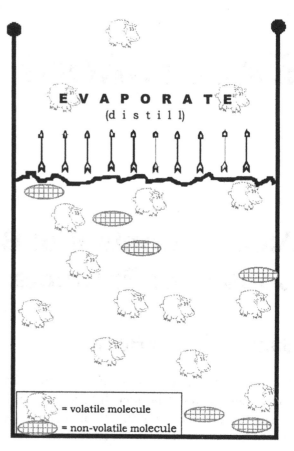

Figure 9.1. Evaporation of a volatile liquid from a beaker.

Using a mixture of water (boiling point 100 °C) and grain alcohol (boiling point 78 °C) as an example, heating of this solution to 78 °C does not result in all the alcohol boiling off at this temperature. Furthermore, it is actually impossible to separate completely just pure alcohol and pure water by distillation. As is so often the case, the more we learn about a subject, the more we find we have left to learn. But the fascinating twists and turns of distillation will have to wait for your next (?) chemistry course!

The method for alcohol analysis described in this experiment is basically the same as that used in the analysis of beer and wine by official government agencies. The alcohol and water do not have to be separated from each other. The purpose of carrying out the distillation is to remove both of the volatile components—the water and all the alcohol—from your sample so that the composition of this liquid mixture may be determined. Direct composition analysis on the original sample itself would be complicated by the presence of especially nonvolatile substances dissolved in the solution such as sugars present in some wines.

Since it is not practical to distill *all* of the liquid in your sample (you should understand this limitation better after performing the distillation yourself), *some* water is added to your sample before distillation, and then distillation is stopped when just this *same volume* of water remains in the flask. As long as the temperature on your thermometer registers close to the boiling point of pure water (100 °C), you should know that by then all the lower boiling alcohol has distilled over into your receiver vessel and that water is the only volatile liquid left in the flask.

There are several ways for determining the percent composition of an alcohol/water mixture. Like procedures followed by official agencies, we will use one to which we already have been introduced—that of density measurement. Pure water will have a density of 1.00 g/cm^3, and pure ethyl alcohol 0.798 g/cm^3, at 20 °C. Mixtures of these two liquids will have densities between these values and, by reference to tables, your measured density can be converted into percent alcohol.

The hydrometer is a device for quickly measuring such densities and is in principle the same way that your car battery fluid can be checked to see if the battery is run down. The more dense a liquid is, the higher the hydrometer will float in it; conversely, the less dense a liquid is, the further down in it the hydrometer will sink. Since all the alcohol from your sample ends up in an equal volume of distillate, the percent alcohol in the distillate will equal the percent alcohol in your sample.

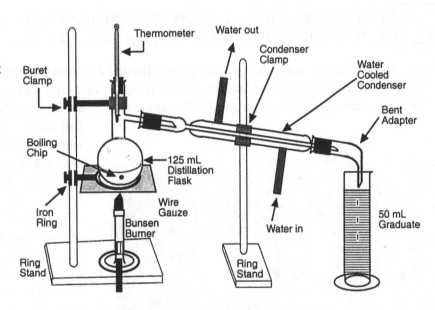

Figure 9.2. Apparatus for a simple distillation.

Liquid densities can also be very accurately determined using a Westphal balance. This balance measures the buoyancy of a calibrated glass "plummet" when immersed in the liquid. This buoyancy effect is dependent upon the density of the unknown liquid. The density can be read directly from the balance itself without calculation.

Procedure

If not already assembled for you when you come to class, set up a distillation apparatus as diagrammed on the previous page and in the picture below. Using clamps and ring stands as necessary, secure both the flask and the condenser. Be careful not to use undue force when inserting your thermometer, especially when using a rubber stopper. Lubricate the hole first with water or glycerine.

1. If your sample is carbonated like beer or sparkling wines, you will need to shake around 60 mL of your sample in a 500 mL Erlenmeyer flask for 5–10 minutes until fizzing subsides before proceeding.

Remove the stopper that holds the thermometer and insert a long stem funnel in the top of the flask. Using a 50 mL graduate, pour in exactly 50 mL of your sample to be analyzed and follow with the addition of 25 mL of water. Remove the funnel and add a boiling chip. If your sample was *carbonated* or is a *mouthwash*, also add a few drops of an anti-foaming agent. Replace the stopper and thermometer.

 ONCE EVERYTHING IS READY, HAVE YOUR SET-UP CHECKED BY THE LAB INSTRUCTOR BEFORE COMMENCING HEATING.

Turn on the water so that a slow stream passes through the condenser and begin heating the flask with a Bunsen burner. Heat strongly at first until a ring of condensing vapors is seen moving up towards the thermometer. (You should also be able to feel the hot vapor ring with your fingers.) Then cut back the heat with the gas supply valve and adjust the heat so that the liquid drips into the graduated cylinder at a rate of about one drop per second. Initially try to adjust the heat by partly closing the gas control valve to the Bunsen burner. It is likely that you still may find it difficult to control the heat and hence the rate of distillation. In this case, slide the wire gauze in and out so as to place *open screen* (more heat) or *white ceramic center* (less heat) between the burner flame and distillation flask. By sliding this gauze around you should be able to find just the right position to give you the desired one drop per second distillation rate.

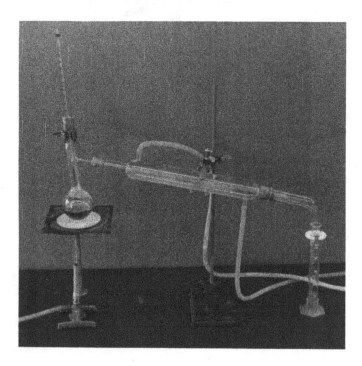

Figure 9.3. Distillation in progress.

2. Note the temperatures of your distillate vapors during the course of the distillation and record your observations. By the time that you have collected 50 mL of distillate, **STOP**—the temperature should have risen to 98–100 °C. (Some samples like perfumes and mouthwashes may yield a cloudy liquid towards the end of the distillation, so don't panic if you observe this.) What *is* important is that you collect exactly 50 ml of liquid—the same volume that you started with.

 3. With the help of your lab instructor, measure the density of your alcohol/water mixture by either using a Westphal balance, or slowly lowering a hydrometer (fragile: see Figure 9.5 on next page!) into the liquid in your graduate. Make sure that the Westphal glass plummet or hydrometer bulb is *not* touching the sides or bottom of the container.

Hydrometers come in different density ranges, so be very careful to read accurately and correctly the proper number off the hydrometer stem. The Westphal balance will have already been standardized against the known density of water by your lab instructor. *Be very careful in the use of this sensitive instrument.*

When you balance the Westphal, make sure that:

> the glass plummet is floating free without touching the walls of the liquid container,

> the glass plummet is completely submerged in the liquid and

> the weights are placed only in numbered balance beam notches (or hung from weights already in a notch).

This balance uses three or four sizes of special weights. After you get the instrument balanced using your liquid, note the *notch number* and *size* for each weight. Each _number_ stands for a *decimal point*, and the weight *size* gives you the *position* of the decimal point. The largest weight represents

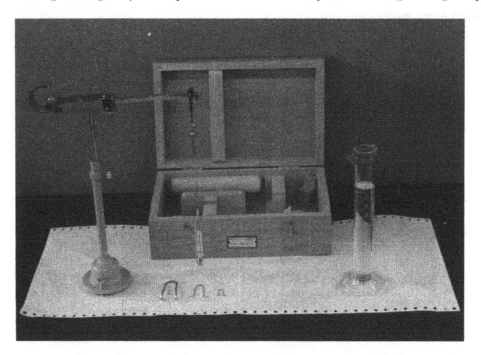

Figure 9.4. Determining density using a Westphal balance.

tenths, next largest hundredths, etc. Three weights would thus permit a density to be read to three decimal places (to thousandths or 0.001 g). Check with your lab instructor for help if you have any questions.

If for any reason you cannot measure the density of your liquid with a Westphal balance or hydrometer, you will have to calculate it from a measurement of the liquid weight and volume as done in Experiment 2: *Going Metric with the Rest of the World.* Note the method of density measurement that you used on the report sheet and record the measured density.

4. Refer to the alcohol/water density tables on the next two pages and determine the alcohol content of your solution. Note on the report sheet whether the composition that you report is percent by *weight* or by *volume* (consult with the lab instructor for help in deciding which is appropriate for your sample). For our purposes, we shall consider density to be the same as the specific gravity.

Figure 9.5. Measuring density with a hydrometer.

Alcohol/Water Density Tables

Specific Gravity	% Alcohol by Volume	% Alcohol by Weight	Specific Gravity	% Alcohol by Volume	% Alcohol by Weight	Specific Gravity	% Alcohol by Volume	% Alcohol by Weight
1.00000	0.00	0.00	0.99417	4.00	3.20	0.98897	8.00	6.42
0.99984	0.10	0.08	0.99403	4.10	3.28	0.98885	8.10	6.50
0.99968	0.20	0.16	0.99390	4.20	3.36	0.98873	8.20	6.58
0.99353	0.30	0.24	0.99376	4.30	3.44	0.98861	8.30	6.67
0.99937	0.40	0.32	0.99363	4.40	3.52	0.98849	8.40	6.75
0.99923	0.50	0.40	0.99349	4.50	3.60	0.98837	8.50	6.83
0.99907	0.60	0.48	0.99335	4.60	3.68	0.98825	8.60	6.91
0.99892	0.70	0.56	0.99322	4.70	3.76	0.98813	8.70	6.99
0.99877	0.80	0.64	0.99308	4.80	3.84	0.98801	8.80	7.07
0.99861	0.90	0.71	0.99295	4.90	3.92	0.98789	8.90	7.15
0.99849	1.00	0.79	0.99281	5.00	4.00	0.98777	9.00	7.23
0.99834	1.10	0.87	0.99268	5.10	4.08	0.98765	9.10	7.31
0.99819	1.20	0.95	0.99255	5.20	4.16	0.98754	9.20	7.39
0.99805	1.30	1.03	0.99241	5.30	4.24	0.98742	9.30	7.48
0.99790	1.40	1.11	0.99228	5.40	4.32	0.98730	9.40	7.56
0.99775	1.50	1.19	0.99215	5.50	4.40	0.98719	9.50	7.64
0.99760	1.60	1.27	0.99202	5.60	4.48	0.98707	9.60	7.72
0.99745	1.70	1.35	0.99189	5.70	4.56	0.98695	9.70	7.80
0.99731	1.80	1.43	0.99175	5.80	4.64	0.98683	9.80	7.88
0.99716	1.90	1.51	0.99162	5.90	4.72	0.98672	9.90	7.96
0.99701	2.00	1.59	0.99149	6.00	4.80	0.98660	10 00	8 04
0.99687	2.10	1.67	0.99136	6.10	4.88	0.98649	10.10	8.12
0.99672	2.20	1.75	0.99123	6.20	4.96	0.98637	10.20	8.20
0.99658	2.30	1.83	0.99111	6.30	5.05	0.98626	10.30	8.29
0.99643	2.40	1.91	0.99098	6.40	5.13	0.98614	10.40	8.37
0.99629	2.50	1.99	0.99085	6.50	5.21	0.98603	10.50	8.45
0.99615	2.60	2.07	0.99072	6.60	5.29	0.98592	10.60	8.53
0.99600	2.70	2.15	0.99059	6.70	5.37	0.98580	10.70	8.61
0.99586	2.80	2.23	0.99047	6.80	5.45	0.98569	10.80	8.70
0.99571	2.90	2.31	0.99034	6.90	5.53	0.98557	10.90	8.78
0.99557	3.00	2.39	0.99021	7.00	5.61	0.98546	11.00	8.86
0.99543	3.10	2 47	0.99009	7.10	5.69	0.98535	11.10	8.94
0.99529	3.20	2.55	0.98996	7.20	5.77	0.98524	11.20	9.02
0.99515	3.30	2.64	0.98984	7.30	5.86	0.98513	11.30	9.11
0.99501	3.40	2.72	0.98971	7.40	5.94	0.98502	11.40	9.19
0.99487	3.50	2.80	0.98959	7.50	6.02	0.98491	11.50	9.27
0.99473	3.60	2.88	0.98947	7.60	6.10	0.98479	11.60	9.35
0.99459	3.70	2.96	0.98934	7.70	6.18	0.98468	11.70	9.43
0.99445	3.80	3.04	0.98922	7.80	6.26	0.98457	11.80	9.51
0.99431	3.90	3.12	0.98909	7.90	6.34	0.98446	11.90	9.59

Alcohol/Water Density Tables
(Continued)

Specific Gravity	% Alcohol by Volume	% Alcohol by Weight	Specific Gravity	% Alcohol by Volume	% Alcohol by Weight	Specific Gravity	% Alcohol by Volume	% Alcohol by Weight
0.98435	12.00	9.67	0.97608	20.00	16.26	0.95185	40.00	33.35
0.98424	12.10	9.75	0.97558	20.50	16.67	0.95107	40.50	33.79
0.98413	12.20	9.83	0.97507	21.00	17.09	0.95028	41.00	34.24
0.98402	12.30	9.92	0.97457	21.50	17.51	0.94948	41.50	34.68
0.98391	12.40	10.00	0.97406	22.00	17.92	0.94868	42.00	35.13
0.98381	12.50	10.08	0.97457	22.50	18.34	0.94786	42.50	35.58
0.98370	12.60	10.16	0.97304	23.00	18.76	0.94701	43.00	36.03
0.98359	12.70	10.24	0.97253	23.50	19.17	0.94620	43.50	36.48
0.98348	12.80	10.33	0.97201	24.00	19.59	0.94536	44.00	36.93
0.98337	12.90	10.41	0.97149	24.50	20.01	0.94450	44.50	37.39
0.98326	13.00	10.49	0.97097	25.00	20.43	0.94364	45.00	37.84
0.98315	13.10	10.57	0.97044	25.50	20.85	0.94276	45.50	38.30
0.98305	13.20	10.65	0.96991	26.00	21.27	0.94188	46.00	38.75
0.98294	13.30	10.74	0.96937	26.50	21.69	0.94098	46.50	39.21
0.98283	13.40	10.82	0.96883	27.00	22.11	0.94008	47.00	39.67
0.98273	13.50	10.90	0.96828	27.50	22.54	0.93916	47.50	40.13
0.98262	13.60	10.98	0.96772	28.00	22.96	0.93824	48.00	40.60
0.98251	13.70	11.06	0.96715	28.50	23.38	0.93730	48.50	41.06
0.98240	13.80	11.15	0.96658	29.00	23.81	0.93636	49.00	41.52
0.98230	13.90	11.23	0.96600	29.50	24.23	0.93540	49.50	41.99
0.98219	14.00	11.31	0.96541	30.00	24.66	0.9344	50.00	——
0.98209	14.10	11 39	0.96481	30.50	25.08	0.9244	55.00	——
0.98198	14.20	11.47	0.96421	31.00	25.51	0.9136	60.00	——
0.98188	14.30	11.56	0.96360	31.50	25.94	0.9021	65.00	——
0 98177	14.40	11.64	0.96298	32.00	26.37	0.8900	70.00	——
0.98167	14.50	11.72	0.96235	32.50	26.80	0.8773	75.00	——
0.98156	14.60	11.80	0.96172	33.00	27.23	0.8639	80.00	——
0.98146	14.70	11.88	0.96108	33.50	27.66	0.8496	85.00	——
0.98135	14.80	11.97	0.96043	34.00	28.09	0.8339	90.00	——
0.98125	14.90	12.05	0.95977	34.50	28.52	0.8161	95.00	——
0.98114	15.00	12.13	0.95910	35.00	28.96	0.7939	100.00	——
0.98063	15.50	12.54	0.95842	35.50	29.38			
0.98011	16.00	12.95	0.95773	36.00	29.83			
0.97960	16.50	13.37	0.95703	36.50	30.26			
0.97909	17.00	13.78	0.95632	37.00	30.70			
0.97859	17.50	14.19	0.95560	37.50	31.14			
0.97808	18.00	14.60	0.95487	38.00	31.58			
0.97758	18.50	15.02	0.95413	38.50	32.03			
0.97708	19.00	15.43	0.95338	39.00	32.46			
0.97658	19.50	15.84	0.95262	39.50	32.90			

Report Sheet Experiment 9

Alcohol Content of Beverages

Date ________ **Section number** ________ **Name** __________________

1. Nature of sample

Product name

Brand____

Where purchased ____

Color____

Smell ____

2. Distillation

(a) Observations during distillation____

____

(b) Distillate characteristics

Color ____

Smell____

(c) Temperature when distillate volume is:

1 mL ________°C.

10 mL ________°C.

20 mL ________°C.

30 mL________°C.

40 mL ________°C.

50 mL ________°C.

3. Density of distillate

    (a) Method used to determine density of liquid
    (Westphal balance, hydrometer or mass-divided-by-volume measurement)

    _____

    (b) Measured density of liquid .... _____ g/cm^3.

4. Percent alcohol (from tables shown in this experiment)

    (a) % alcohol in distillate = % alcohol in sample .... _____ %.
    (specify % by *weight* or % by *volume*)

    (b) % alcohol in sample from label .... _____ %.
    (specify % by weight or % by volume)

5. Comments and conclusions on the experiment

Alcohol Content of Beverages

Date _____   **Section number** _____   **Name** _____

1. What do your temperature readings suggest about the effectiveness of your skill in separating out pure alcohol? Explain. What temperature readings would you have seen if all the pure alcohol had distilled off by itself at the beginning of the distillation?

2. How effective was the distillation in removing colored materials from your sample? What does this suggest about the *volatility* of these materials? (Compare distillate color to the original sample.)

3. Now that you have performed this experiment, explain why it would have been impractical to distill <u>all</u> of the liquid from your 50 mL sample over into your receiver. In other words, why was it necessary to add the 25 mL of water to your flask before commencing the distillation?

4. The moonshiner makes *white lightning* in a big vat, which is later subjected to a distillation. What exactly does this distillation accomplish? Why is the "distilled product" better than the "vat product," or is it? Be specific.

5. Why do you think products like colognes and mouthwashes contain alcohol?

Think, Speculate, Reflect and Ponder

6. How does the process of distillation relate to the fact that rain is usually substantially lower in dissolved chemicals than the surface water it falls into?

7. With the help of your lecture text or other source, list two important advantages of using "gasohol" motor fuels (gasoline containing alcohol).

    (a)

    (b)

Experiment 10

Why Is Water Harder Than Iron?

Sample From Home

Bring 200 mL of a water sample from a tap, well, lake, etc.

Objectives

The chemical differences between soaps and detergents will be demonstrated and their respective advantages and disadvantages emphasized. The technique of *titration* will be used to compare quantitatively the effectiveness of both a soap and a common detergent in soft and hard water. Quantitative information using a standard hard water will enable the specific hardness value for an unknown sample from home to be determined.

Background

You are no doubt ready to argue that certainly water is *not* harder than iron. Ah, but it *is* if one defines hardness based on the concentration of calcium and magnesium ions present. In that case, most samples of water would, by such a definition, indeed be "harder" than pure iron because calcium and magnesium ions are commonly found in water supplies. Admittedly some tricky semantics have been thrown at you, and in truth it must be acknowledged that such special hardness applies only to samples of water and not iron. This type of hardness has thus more to do with the chemical, rather than physical, characteristics of water. All official water analyses include data for the water hardness since it plays an important and unwelcome economic role by causing

> chemical interference in some industrial processes;
> scale formation inside home and industrial plumbing installations;
> excessive consumption of soap.

These disadvantages of hard water are due primarily to the presence of calcium and magnesium ions, although the official analysis procedure determines the total amount of all *alkaline earth* (group II) metal ions in the periodic table. While calcium may be found in sewage and industrial effluents, the *alkaline earth* metals commonly found in water usually come from their soluble salts which have been leached out into ground water from surrounding rock strata.

Limestone and dolomite are two such prevalent rocks found in the earth's crust and consist of calcium and magnesium carbonates. These carbonate minerals are almost completely insoluble in water, but contact with acids which are present in all natural waters (commonly due to dissolved carbon dioxide) convert these *insoluble* carbonates into the much more *soluble* bicarbonates. Calcium bicarbonate, for example, is 30 times more soluble than calcium carbonate and is produced according to the reaction:

$$CaCO_3 \quad + \quad CO_2 \quad + \quad H_2O \quad \longrightarrow \quad Ca(HCO_3)_2$$

calcium carbonate   carbon dioxide   water   calcium bicarbonate
rock (insoluble)   (more soluble)

Magnesium carbonate behaves similarly to calcium bicarbonate. These bicarbonates are especially troublesome in boilers and hot water tanks; heating water containing these salts produces an insoluble precipitate (scale). Not only can these deposits seriously restrict or stop the flow of water as the scale builds up, the scale layer also seriously reduces the heating efficiency of the apparatus by acting as a heat transfer barrier. Heat essentially reverses the original reaction in which the bicarbonate was formed.

$$Ca(HCO_3)_2 \quad \xrightarrow{\text{Heat}} \quad CaCO_3 \quad + \quad CO_2 \quad + \quad H_2O$$

calcium bicarbonate   calcium carbonate   carbon dioxide   water
(soluble)   (insoluble scale)

Because heating and boiling water containing calcium and magnesium bicarbonates can remove these alkaline earth ions by causing them to precipitate, the term *temporary hardness* is sometimes applied in such cases. If other calcium/magnesium minerals such as sulfates ($CaSO_4$ and $MgSO_4$) are present, they are unaffected by heating and hence this kind of hardness is referred to as *permanent hardness*. Official analyses of water for total hardness (as well as the method you will use in this experiment), however, measure *all* the calcium, magnesium and other *alkaline earth* minerals present—bicarbonates, sulfates, etc.

Physiologically, hardness does not appear to be detrimental. In fact, hard water decreases the sensitivity of fish to toxic metals, and experiments with calves and chicks have indicated that those supplied with hard water develop somewhat better than those supplied with distilled water containing none of these elements. And because of its mineral content, hard water is usually superior to soft water for irrigation.

Now that we have looked a little at the nature of *hard* water, we are in a better position to discuss the particular problem associated with hard water which is examined in this experiment—that of its effect on *soaps* and *detergents* and on excessive soap consumption. All soaps and detergents, have a basic feature in common—they consist of very long molecules. One end of this long molecule consists of a nonpolar hydrocarbon chain which tends to dissolve in oil and grease (which are both nonpolar hydrocarbons themselves); the other end is very polar or even ionic and likes to dissolve in water (a very polar liquid and good solvent for ionic substances).

An axiom familiar to chemists is "Like dissolves like." When a soap or detergent is shaken with oil and water, the long soap/detergent molecule binds together tiny oil and water droplets by virtue of the different solubilities of each end of its long molecule; each end dissolves in the substance in which it is most "like."

dissolved in water   *dissolved in oil droplet*

polar/ionic end   *hydrocarbon chain of molecule*

You may have heard that oil and water don't mix. In this case, however, billions of these oil+water droplet combinations produce what is called an emulsion—a relatively stable mixture of oil and water. Once formed, this emulsion can be simply flushed away with water along with dirt particles and the cleanser has thus done its job.

The large scale commercial preparation of *soap* today is basically the same as was practiced centuries ago with tallow and ashes—the splitting apart of animal or plant fats and oils with lye:

a fat or oil   sodium hydroxide (lye)   glycerol (glycerin)   sodium salts of three fatty acids ("soaps)

The shaded box at the far right denotes that part of the product molecules which are *oil* soluble and the gray hatched box those parts that are *water* soluble. Because of the non-uniformity typical of natural products like fats, the hydrocarbon chains (symbol R) present will differ in length as indicated by R_1, R_2 and R_3. The best soaps are those in which the "R" groups contain between nine and seventeen carbon atoms. With fewer than nine carbon atoms, insufficient oil solubility and emulsification occur, while having over seventeen carbons makes the soap too insoluble in water to be effective.

Slight modifications in the chemical and physical makeup of *soaps* can give rise to a wide variety of familiar products:

> floating soaps (with air beaten in),
> soft or liquid soaps (using potassium instead of sodium salts of the fatty acids),
> castile soaps (using olive oil—a liquid fat—to make soap),
> transparent soaps (alcohol added to the soap mix), and
> perfumed and germicidal soaps (appropriate chemical dissolved in soap).

During World War II, people were asked to buy victory bonds, start a victory garden and recycle tin foil which kids made into foil balls from cigarette package liners (a Lucky Strike package offered a bonus that let you sock that bully without fear of retaliation). But also saved were fats produced from cooking which were taken to the local grocery that sent them on to a central chemical processing plant. The desired product in this case was not mainly soap but the *glycerin*, a necessary ingredient for making nitroglycerin used in dynamite. (Alfred Nobel's discovery of a "safe" way to handle nitroglycerin in the form of dynamite led to the establishment of the prestigious Nobel prizes.)

But what's so wrong with *soaps* that we need *detergents*, then? Acidic waters can precipitate some of the soap and thereby inactivate it. However a much greater problem lies in the fact that soap molecules will combine with certain metal ions such as the *alkaline earths* (and even certain heavy metals like iron ions) to give an insoluble precipitate. The precipitated *soap* is what causes the fa-

$$2R \overset{\overset{\displaystyle O}{\|}}{-C} -O^- \; Na^+ \;\; + \;\; Ca^{+2} \;\; \longrightarrow \;\; \left(R \overset{\overset{\displaystyle O}{\|}}{-C} -O^-\right)_2 Ca^{+2} \;\; + \;\; 2Na^{+1}$$

| a soap (soluble) | calcium ions (soluble) | calcium salt of soap (insoluble crud) | sodium ions (soluble) |

miliar scum of "bathtub ring" and results not only in a clean-up mess, but also a significant waste of soap. Before any cleansing action can occur, enough soap has to be added to neutralize both water acidity as well as neutralize (precipitate out) all ions like calcium which are present in the water. Until the advent of *detergents*, the only alternatives to such scum and wasted soap were:

> use rain water for washing;
> add other chemicals to try to tie up (inactivate) the acidity or calcium-like ions;
> install a Zeolite type ion exchange resin in the water supply system.
  (When water flows over such a resin, the "hard" calcium/magnesium ions are exchanged for "soft" sodium ions, the resin thus acting as a water softener.)

Then the 1950s marked the entrance of *detergents*. Like most carbon containing materials we use, these compounds are derived ultimately from petroleum through chemical synthesis. Some common types of detergents are:

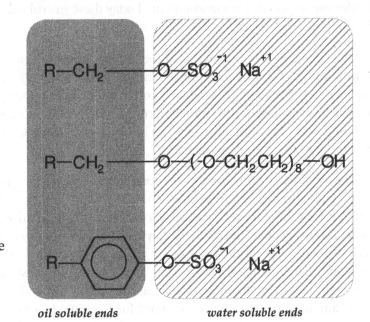

1. Sodium alkyl sulfates (basic solutions)

2. Ethoxylates (non-ionic detergents; neutral solutions)

3. Sodium alkylbenzene sulfonates (basic solutions)

oil soluble ends   *water soluble ends*

Of these three types, the sodium alkylbenzenesulfonates (ABS) detergents are the most widely used.

The big advantage common to all these detergents is the very one lacking in all *soaps*—the ability to clean in cold and hard water without being precipitated by acids or, especially, calcium and magnesium type ions. In fact, the calcium and barium salts of these compounds are used as detergent additives in engine oils. Furthermore, the cleaning action of detergents does *not* depend upon the formation of foam, which makes possible their use as low-foaming cleansers necessary for automatic dish and clothes washers.

Prior to 1965, however, detergents suffered one very big disadvantage—they were not *biodegradable* like soaps. Because of this, literally mountains of stable foam were not uncommon sights along waterways. Chemists discovered that this lack of biodegradability was due to the nature of the hydrocarbon chain "R." But if this "R" group were made into an essentially straight carbon chain without

$$CH_3-\underset{\underset{CH_3}{|}}{CH}-CH_2-\underset{\underset{CH_3}{|}}{CH}-CH_2-\underset{\underset{CH_3}{|}}{CH}-CH_2-\underset{\underset{CH_3}{|}}{CH}-\underset{\bigcirc}{}-SO_3^{-1}\ Na^{+1}$$

non-biodegradable ABS detergent

$$CH_3-CH_2-CH_2-CH_2-CH_2-CH_2-CH_2-CH_2-CH_2-CH_2-\underset{\underset{CH_3}{|}}{CH}-\underset{\bigcirc}{}-SO_3^{-1}\ Na^{+1}$$

biodegradable ABS detergent

any side branches, bacteria could digest it and would indeed chew the detergent up so it would not remain after use to pollute the environment. Today these microbially chewable detergents have replaced their earlier *non-biodegradable* counterparts.

Commercial detergent products may also contain the following: water softeners (formerly phosphates) to inactivate Ca, Mg and similar ions; optical brighteners (special light absorbing chemicals which make clothes appear brighter); perfumes for sales appeal; bleaches to increase the whiteness of clothes; enzymes to remove protein-based soil and stains. (Good articles chronicling the development of soaps and detergents can be found in *Today's Chemist at Work* October, 1996, and *Chemical & Engineering News* January 27, 1997.)

An official procedure for determining the total hardness of water calls for the *titration* of the *alkaline earth metals* with a reagent called disodium dihydrogen ethylenediamine tetraacetate (Na_2EDTA for short!) and uses a special color indicator to detect the *end point*. However, we can obtain reasonably accurate results by simply using a soap solution as the titration reagent, taking advantage of soap's normal disadvantage of being precipitated by alkaline earth ions. As soon as all the alkaline earth "hardness ions" (mainly calcium and magnesium) have been thus precipitated, the soap will start to foam upon shaking and that is our end point. Such a *quantitative* measure of the amount of soap solution required for foaming, together with a comparison to a standard hard water solution, will permit you to determine the actual hardness of an unknown water sample.

Whatever the hardness is due to, it is normally reported in units of milligrams of $CaCO_3$ per liter or, what is the same, parts of calcium carbonate per 1 million parts of water (ppm $CaCO_3$). The quantitative water hardness can be related to a classification of water type using the following table:

Hardness (ppm $CaCO_3$)	Classification
0–60	Soft
61–120	Moderately Hard
121–180	Hard
Over 180	Very Hard

Table 10.1 Water type classification.

Procedure

Set up two burettes for a titration at your bench station using a double burette clamp. Refer to Experiment 13: *Vitamin C in Your Diet* regarding the preparation of a burette and the description of this procedure. You will also need four 250 mL Erlenmeyer flasks with stoppers, all of which have been thoroughly cleaned with soap and water and rinsed with distilled water. Fill one burette with 0.5% detergent solution and the other with 0.5% soap solution, to a level about 1 inch above the 0.00 mL line. Label the burettes so you will not mix them up later.

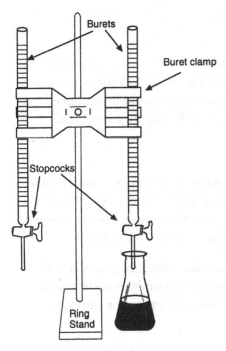

Figure 10.1. A titration assembly.

Open the stopcock valve and allow the liquid to fill the burette tip and wash out any air bubbles. When the meniscus level (recall Experiment 2: *Going Metric With the Rest of the World*) drops even with the 0.00 mL mark, shut off the stopcock and discard the liquid that has drained out. Examine the graduations on your burette carefully—both their numerical equivalent and the direction the numbers are increasing—to be sure you will read the meniscus levels correctly during the rest of the experiment. Your lab instructor can review the operation and reading of a burette before you actually begin.

Note: This experiment can be completed more quickly if the titrations are divided up between partners.

Detergent Titrations

1. Using your 50 mL graduated cylinder, pour 50 mL of distilled water into each of two clean 250 mL Erlenmeyer flasks (you will not need all *four* flasks until Parts 2(b) and 2(c)) and proceed to titrate each of your duplicate water samples with the detergent solution. Your endpoint in this titration will be the minimum amount of detergent solution causing the formation of a stable foam covering about ¼ of the liquid surface which persists for at least 20–30 seconds. Stopper your flask and briefly shake vigorously after addition of each 0.1 mL (2 drops) of detergent solution.

The evaluation of the *stable foam endpoint* is somewhat subjective, so don't unnecessarily worry about what is "enough" foam. What is more important is simply trying to get the *same* foam endpoint (whatever you have accepted) for *all* samples. Your lab instructor can demonstrate what such an endpoint looks like if you continue to have trouble.

(a) Record data for the volume of detergent solution necessary to titrate the 50 mL of distilled ("soft") water to the nearest 0.1 mL. Repeat with the second 50 mL sample of distilled water. The two volumes needed to reach the end point for these duplicate samples should agree within 0.1 mL of each other. If not, run a third trial on another 50 mL sample of distilled water.

(b) Rinse out your flasks thoroughly with tap water, rinse with distilled water and fill each with 50 mL of the standard hard water solution. Repeat the titration procedure as in 1(a) and record the volume data for detergent solution required to titrate each of your duplicate 50 mL standard hard water samples. As before, if the two detergent volumes do not agree within 0.1 mL, run a third sample. All of your detergent solution volumes required for each titration in both parts 1(a) and 1(b) should measure under 0.5 mL.

Soap Titrations

2. Again using your 50 mL graduate, pour 50 mL of **distilled water** into each of two clean 250 mL Erlenmeyer flasks and titrate as in Part 1 (a), only this time shake after each 0.25 mL portion (about 5 drops) of **soap** solution is added. Less than 1 mL of soap solution should be required, and your duplicate titrations should agree within 0.25 mL of each other. If not, run a third trial.

   (a) Record all data on the report sheet.

SAVE your best titrated sample from 2 (a)—the one whose foam layer you deem represents the best end point—for comparison with the following titrations in Parts 2 (b) and 2 (c). This soap/distilled water solution represents your *reference foam* condition, to which you can compare your foams obtained in subsequent titrations in order to be able to obtain good quantitative information.

Again, remember: the evaluation of the *stable foam endpoint* is somewhat subjective, so don't unnecessarily worry about what is "enough" foam. What is more important is simply trying to get the *same* foam endpoint (whatever you have accepted) for *all* samples. Your lab instructor can help you decide what such an endpoint looks like if you are uncertain.

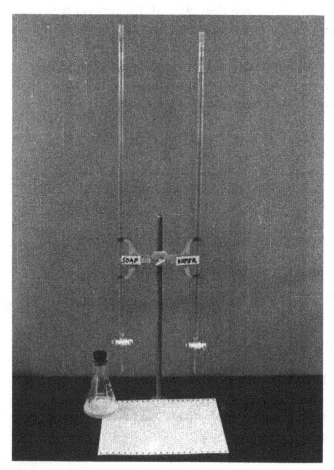

Figure 10.2. Soap solution and water in separate burettes ready for a titration.

Pour 50 mL of the **standard hard water** solution into each of your three remaining cleaned 250 Erlenmeyer flasks and titrate as in 2 (a) with the **soap** solution. You will find it fastest to titrate the first sample rapidly, shaking only after each 1 mL portion of soap solution is added. Your rough results from this titration should tell you the approximate volume of soap solution required to the nearest 1 mL.

(b) Now proceed to titrate the other two samples of **standard hard water** by adding the soap solution rapidly at first until you get to within 1 mL of the estimated end point. Then slow down and add the soap in 0.25 mL increments, pausing after each addition to shake and compare the foam produced in the sample you are presently titrating to that in your freshly shaken *reference foam* flask from step 2. When the two foam layers match up, that is your end point. As in step 2, duplicate titrations should agree within 0.25 mL. If not, run a third trial. Record all data on the report sheet.

(c) Rinse well, lastly with distilled water, the three 250 mL Erlenmeyer flasks used in step 2(b), but *don't throw out* your reference foam solution. Fill each flask with 50 mL of your own **unknown water** sample. Repeat exactly the procedure described in 2(b) and titrate with the **soap** solution. Record all your data.

The calculations which follow have you compute the *titer* of the soap solution—much like the titer described for the indophenol solution in Experiment 13: *Vitamin C in Your Diet.* The titer is simply a conversion factor which relates, in this case, a specific volume of soap solution with a specific amount of water hardness. This titer will permit you to convert from units of mL of soap into ppm $CaCO_3$ hardness for your unknown, just like you carry out a conversion from inches into centimeters. The necessary set-ups appear on the report sheet.

Report Sheet

Water "Harder Than Iron"

Date _____ **Section number** _____   **Name** _____

Data

1. Detergent Titrations	Trial I	Trial II	Trial III (Optional)
(a) Distilled water final burette reading	_____	_____	_____
Initial burette reading	_____	_____	_____
Net volume of detergent solution (to nearest 0.1 mL)	_____	_____	_____

Average of two closest trials   _____mL detergent per 50 mL of distilled water.

	Trial I	Trial II	Trial III
(b) Standard hard water final reading	_____	_____	_____
Initial burette reading	_____	_____	_____
Net volume of detergent solution (to nearest 0.1 mL)	_____	_____	_____

Average of two closest trials   _____mL detergent per 50 mL std. hard water.

2. Soap Titrations	Trial I	Trial II	Trial III (Optional)
(a) Distilled water final burette reading	_____	_____	_____
Initial burette reading	_____	_____	_____
Net volume of soap solution (to nearest 0.25 mL)	_____	_____	_____

Average of two closest trials   _____mL soap per 50 mL of distilled water.

	Trial I	Trial II	Trial III (Optional)
(b) Standard hard water final reading	————	————	————
Initial burette reading	————	————	————
Net volume soap solution (to nearest 0.25 mL)	————	————	————

Average of two closest trials _____mL soap
per 50 mL of std. hard water.

	Trial I	Trial II	Trial III (Optional)
(c) Unknown hard water final reading (your sample from home)	————	————	————
Initial burette reading	————	————	————
Net vol. of soap solution (to nearest 0.25 mL)	————	————	————

Average of two closest trials _____mL soap
per 50 mL of unk. hard water.

Calculations

3. Titer of soap solution

(a) Volume of soap required to titrate standard hard water
(final line in 2(b)) _____mL.

(b) Volume of soap required just to make a foam
in distilled water (the "blank") (final line in 2(a)) _____mL.

(c) Net soap volume required to titrate just the hardness
in 50 mL of the 100 ppm standard hard water
(subtract line 3(b) from line 3(a)) = _____mL.

(d) "Titer" of soap solution—parts per million
hardness per 1 mL of soap solution
(divide 100 ppm hardness by line 3 (c)) _____ppm.

(The hardness concentration of the *standard hard water* is 100 ppm.)

4. Hardness of unknown water sample

    (a) Volume of soap required to titrate unknown water
    (final line in 2(c)) .... _____mL.

    (b) Volume of soap required just to make a foam
    in distilled water (the "blank" in line 3(b) .... _____mL.

    (c) Net soap volume required to titrate just the hardness
    in 50 mL of the unknown water sample
    (subtract line 4(b) from 4(a)) .... _____mL.

    (d) Quantitative hardness of unknown water
    (multiply line 4(c) times line 3(d)) .... _____ppm.

Interpretation

5. Classification of hardness
    (from table in lab background discussion) _____.

6. Reference information on unknown

    (a) Geographical location where water sample taken _____.

    (b) Water district from which unknown water sample taken _____.

    (c) Official water district figure for total hardness
    (telephone to find out hardness) .... _____ppm.

7. Comments and conclusions on experiment

Water "Harder Than Iron"

Date _____   **Section number** _____   **Name** _____

1. Explain why the volumes of *detergent* solution required to titrate soft and hard water in l(a) and l(b) on the report sheet are so *similar*, whereas the corresponding volumes of *soap* solution in 2(a) and 2(b) are so *different.*

2. What is the source of the hard water ions found in water from ground water sources?

3. What former major environmental problem associated with the use of detergents has been largely overcome by chemical modification of the molecular structure of the detergent molecules? (Hint: check background discussion to experiment.)

4. List two advantages that detergents have over soaps.
   (a)

   (b)

Think, Speculate, Reflect and Ponder

5. In the past, phosphates were usually added to many cleaners for "extra" cleaning power. What problems do phosphates pose for the environment? (Consult lecture text for help, if necessary)

6. Did the Romans, quite successful plumbers in their own right, have problems with hard water? Why?

Experiment 11

The Staff of Life and Chlorine
(salt & water in bread)

Samples From Home

Bring one slice of fresh bread or similar food product suspected of containing salt. Crackers (6) or pretzels (3 large) may also be used. Bring them in a sealed plastic bag to prevent drying out prior to analysis.

Objectives

You will quantitatively determine the amount of moisture and salt in your bread or related food product and be introduced to the techniques of ignition, precipitation, filtration, and drying. Some basic chemical reactions will also be illustrated.

Background

The representative nonmetal chlorine can be found not only in things that feed us, but also in bleaches, insecticides, and poison war gases. It is indeed a versatile element which can either sustain or kill life, depending upon its chemical form and concentration.

You will utilize some simple chemical reactions to determine the amount of chlorine in a sample of your choosing. This chlorine will be present in the form of chloride ion—most probably due to sodium chloride (common table salt). After drying your sample to determine the amount of moisture present (moisture you—or someone—paid for), the product is ignited to burn away most of the organic compounds which would otherwise plug up the filter paper when the filtration step is attempted. The burnt charcoal residue containing minerals (mainly salt) is extracted with water to dissolve out the salt and the mixture filtered to separate the salt solution from the insoluble charcoal residue. Since some of your salt containing filtrate remains with the charcoal and filter paper, a correction is made for this by a simple calculation shown at the end of the experiment.

When silver nitrate is added to the salt solution filtrate, the salt (NaCl) reacts according to the equation:

$$AgNO_3 \; + \; NaCl \; \longrightarrow \; NaNO_3 \; + \; AgCl$$

| silver nitrate (soluble) | sodium chloride (soluble) | sodium nitrate (soluble) | silver chloride (insoluble precipitate) |

The precipitate formed from treatment with silver nitrate is filtered from the solution, dried and its weight determined. Note that its color changes during handling. Why? (HINT: Silver salts are basic ingredients in light sensitive photographic film emulsions.)

Although other ions besides chloride can also be precipitated by silver nitrate, the residue you see left on your filter paper is mostly silver chloride, AgCl. Since the amount of silver chloride will be small (perhaps 0.20 to 0.40 g), you will need to weigh your precipitate carefully if you expect to get meaningful results. And again, don't forget to *zero your balance* before you use it and *use the same balance* for all your weighings. Expect typical salt concentrations to fall in the range of 1–2% NaCl.

Throughout history bread has been referred to as *"The Staff of Life"* and in some form continues to be one of the world's most important foods for filling hungry stomachs. Do you believe that "bread is bread," or is there really a difference? Even lifting a loaf off the grocery store shelf suggests that there really is indeed a difference. The "enriched" white may feel like an air balloon, while the multi-grain wheat feels like a rock! You should in any case find the following article on *"The Story of Bread"* very interesting.

The Story of Bread

In its natural state, freshly milled wheat flour has a pale yellowish tint, and if allowed to age naturally it slowly will turn white and undergo an aging process that improves its baking qualities. Impatient with natural aging, mass-producers of bread discovered that this process could be accelerated through the addition of chemical oxidizing agents. To speed the production line from mill to oven, industry now adds bleaching agents to flour. Benzoyl peroxide helps make flour "whiter than white" without affecting the baking properties. Others both bleach and age the flour.

To further improve the appearance of bread, if not the quality, the baking industry will add oxidizing substances. These "bread improvers" contain inorganic salts, such as ammonium and calcium phosphates, to serve as yeast food and dough conditioners. Mineral salts may be added to stabilize the gas-retaining properties of the flour gluten, and cyanide or chlorinated organic compounds may be employed in fumigation of the resulting flour in storage.

Before wheat seeds are planted, they are most likely treated for plant disease protection, and the soil in which the wheat is grown is probably infused with fertilizers and no doubt contains the built-up residue of numerous complex and poisonous insecticides, pesticides, herbicides and fungicides used on previous crops grown in the same soil. After harvest and during the storage period prior to milling, grains are subjected to another exposure to the organic poisons used to prevent plant disease or rot and insect and rodent damage. One hopes that these chemicals do not migrate with the wheat grain to the flour.

In the baking industry, a variety of chemical additives are used to produce carbon dioxide, which causes the dough to be light and porous.

The water which is added to the flour may have been chemically purified by means of alum, serta ash, copper sulfate and chlorine. Sugar or dextrose is added to the mixture and in its refining lime, sulfur dioxide, phosphates and charcoal are used. The salt may contain iodide and agents such as calcium and magnesium carbonates to promote free running and prevent caking. The yeast used is treated and fed with ammonium salts. The shortening is a refined, bleached, deodorized product which may contain traces of nickel, be glycerinated and contain anti-oxidants in combination with citric acid, ascorbic acid and phosphoric acid.

The basic ingredient of bread is milled wheat flour, a grain which is composed of three principle parts: the outer shell or husk; the endosperm or kernel; and the germ from which the grain reproduces itself. When the grain is planted, the husk protects the seed while it germinates and the endosperm—mostly carbohydrate—feeds the germ until it gets a foothold and takes nutrients from the earth and air. When the mature grain goes to the mill to be converted to flour, the milling process removes the outer husk and the germ, leaving the endosperm. What emerges is pure starch which, when mixed with water, becomes an easily shaped paste-flour.

The removed bran is sold as feed for animals; the wheat germ is sold, ironically, as a food supplement for human beings and animals. These discarded parts contain the nutrients essential to human health and life. The husk is composed of minerals and vitamins, including the essential B vitamins. The wheat germ, along with its high protein and mineral content, is rich in Vitamin E and a complete Vitamin B complex.

Since the original nutrients in the wheat grain have been removed leaving almost pure carbohydrate, there must be a reintroduction of some portion of the original vitamins and minerals if the end product, bread, is to be sold to the consumer with nutrient value. Thus the fiction of "enriched bread" and the addition of percentages of the adult daily minimum requirements of thiamine, riboflavin, niacin and iron. The milling process removes the B vitamins, several minerals, Vitamin E and protein from the wheat grain. "Enrichment" is a government requirement to replace a percentage of three discarded vitamins and one mineral.

In order to achieve the soft, spongy, foam-like texture of today's commercial white bread, large amounts of chemical emulsifying agents are employed to turn the dough into a mass of uniform tiny air bubbles glued together by the other ingredients. These emulsifiers or surface active agents make it easier for the baking industry to machine the bread dough, increase the volume to weight ratio by incorporating more air into the dough, and slow down the foaming rate. None of the emulsifier chemical additives add to the nutritional qualities of the bread and, in the case of polyoxyethylene fatty esters, enable the baking company to reduce the amount of shortening and milk solids in a loaf of bread.

Besides using calcium propionate to extend the shelf life of bread and retard spoilage, sodium diacetate, sodium propionate, acetic acid, lactic acid and monocalcium phosphate are effective in controlling the growth of bacteria or molds.

The wrapper surrounding a loaf of bread does not list the large number of intentionally or unintentionally added chemical substances included in each slice. Terms such as "dough conditioners", "yeast nutrients" and "enriched" do not indicate the kind, the number or purpose for these added compounds. Peek-a-boo vitamins and minerals, emulsifying agents to prevent all this and more are parts of the chemical creation of an industrial product called bread.

A corollary result of this industrialization of bread production is that vital nutrients are lost in the process and only partially replaced, and a host of extraneous chemicals, otherwise foreign to the human system, are ingested with the consumption of each slice of bread by the public. The consumer, uninformed of the extent and nature of these productional changes and chemical additions, has only taste, appearance and price as criteria to judge the product on the grocery shelf.

The wrapper label fails to indicate to the consumer that "Enriched Bread," unlike the rose, which by any other name is still the same, is not bread.

Procedure

1. NOTE: All glassware used in this experiment should be rinsed with distilled water before use.

(a) Weigh one slice of bread (or 20 g of some similar food item thought to contain salt, such as 6 crackers or 3 large pretzels). Record this and all subsequent weights to the nearest 0.01 g.

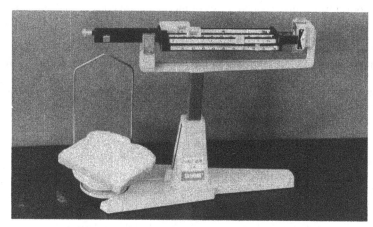

Figure 11.1. Weighing a bread slice on a triple beam balance.

(b) Dry your sample at 180 °C in an oven until *snappy crisp* like rock hard Zwieback (Melba) toast. This may take up to 10–15 minutes depending upon the nature of your particular sample. Reweigh and record data on the report sheet.

Figure 11.2. Drying bread in a drying oven.

Figure 11.3. Breaking dried bread into an evaporating dish.

Figure 11.4. Roasting bread with a Meeker burner.

burner (or use two Bunsen burners) until burning and smoking ceases and charred chunks remain in the dish (10–15 minutes). You may need to scrunch the charring residue down into the dish during heating to keep the burning sample in contact with the hot dish. The dish bottom should have glowed a dull red during the latter part of this heating time.

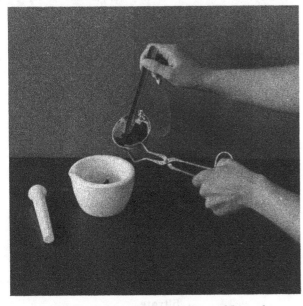

Figure 11.6. Placing charred bread into a mortar.

(c) and (d) Compute the weight loss and percent moisture in your sample.

2. Break up your sample into small pieces and crunch into an evaporating dish. Place the dish onto a small iron ring mounted on a ring stand IN A HOOD and close the hood overhead door. Heat strongly with a Meeker

Figure 11.5. Charred bread residue.

(a) Record your observations from heating.

Clean a mortar and pestle with brush, soap and water. Rinse with distilled water. Grab the evaporating dish (it need not be cool) with tongs and use a scoopula to scrape out the charred chunks of residue into the mortar. Scrunch this residue into small granules with the pestle, add 50 mL of distilled water using a graduated cylinder, and stir for a couple of minutes.

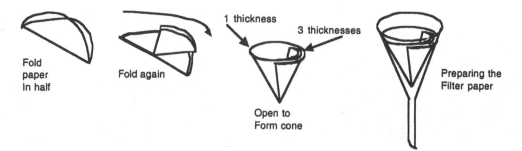

Figure 11.7. How to fold a filter paper to make a cone.

Place a long-stemmed funnel into a ring mounted on a ring stand at your desk station. Fit it with a filter cone made from 12.5 cm paper and place a 50 mL graduate directly underneath.

Dump the liquid/solid contents of your mortar (called a *slurry*) all at once into a 400 mL beaker. Pour your liquid slurry in the beaker into your filter paper cone, taking care not to get the liquid level above the top edge of the paper cone. Collect 25–35 mL of clear filtrate, which should take 5–10 minutes. (The remaining unfiltered liquid and black solid left on the filter paper may be discarded.)

(b) Record the volume of the filtrate collected.

3. Pour your filtrate into a 125 mL Erlenmeyer flask. Acidify by adding 5 mL of dilute nitric acid ($6M$ HNO_3) and then add 15 mL of silver nitrate solution ($0.2M$ $AgNO_3$) to the flask.

(a) What do you observe?

Figure 11.8. Filtering water extract from bread residue.

Figure 11.9. Adding silver nitrate to bread filtrate.

A white solid (*precipitate*) formed in this step indicates the probable presence of chloride ions. For confirmation and to obtain a *quantitative* determination (measurement of *how much*), proceed as follows: Using a burette (utility) clamp, attach the flask to a vertical rod so it rests directly on top of a wire gauze screen held by an iron ring. Heat the flask contents to boiling with a Bunsen burner.

Continue boiling cautiously (watch out for excessive frothing) for one minute. Stop heating with the Bunsen burner and allow the precipitate to settle for a couple of minutes. If you have done things properly (and they must be for success here), the initially *milky* solution in the Erlenmeyer flask should now have a precipitate at the bottom with a mostly *clear* liquid (the *supernatant*) on top. If your solution is still milky at this point, add a few more milliliters of $6M$ HNO_3 and boil for five minutes or until the precipitate coagulates and settles when the heating is stopped.

(b) To check that all of the chloride ions have reacted to form a precipitate, add a few drops

Figure 11.10. Heating liquid containing silver chloride precipitate.

of $0.2M$ $AgNO_3$ to the supernatant (clear liquid) with a medicine dropper. No additional cloudiness should appear. If you do observe cloudiness, add 10 mL of additional $0.2M$ $AgNO_3$ and repeat the boiling procedure.

4. Carefully (steady hand!) pour off (*decant*) most of the clear upper liquid layer (*supernatant)* from the precipitate and discard this liquid. Place a filter funnel inside your iron ring and fit it with a paper cone (Whatman #1, 9 cm). Place a beaker underneath to catch liquid passing through the filter paper. Slop/pour the small amount of solid-liquid slurry from your Erlenmeyer flask into the paper cone.

Figure 11.11. Decanting clear liquid from silver chloride precipitate.

Wash the remaining bits of precipitate out of your flask and into the filter paper with two small (10-15 mL) portions of distilled water. Discard the filtrate. If squirt bottles of distilled water are available, hold the flask upside down over the filter paper cone and use a distilled water stream to wash out the remaining bits of precipitate.

When all the liquid has drained through, CAREFULLY—without tearing—lift the filter paper cone out of the funnel and gently place it upright nestled in the mouth of a 50 mL beaker. (You can put your initials on the frosted white spot on the side of the beaker with a pencil for easy identification later.) Set the beaker in an oven to dry at 160 °C until the paper itself feels dry. This should take around ten minutes. (Gently squeeze the bottom tip of the filter paper cone with your fingers. If any dampness is felt, return paper to the oven for five

Figure 11.12. Washing precipitate into filter paper using a squirt bottle.

minutes more.) Remove the beaker from the oven using tongs or paper towels and allow it to cool to room temperature.

(a) Weigh and record to the nearest 0.01 g the weight of the filter paper and precipitate.

(b) Weigh and record the weight of a new, unused filter paper circle accurately to the nearest 0.01 g and record.

(c) Calculate the net weight of your silver chloride precipitate and record on the report sheet.

(d) You should by now, if not earlier, be able to see some changes in the appearance of your precipitate since it was first formed in step 3(a). Record your observations.

Figure 11.13. Heating silver chloride precipitate in a drying oven.

5. Complete the calculations called for on the report sheet.

Chemistry math usually involves, as a minimum, a proficiency in algebra and use of conversion factors. However, all calculations in this manual can be performed by simply following the set-up directions on the report sheets.

The Staff of Life

Date _____　　　**Section number** _____　　　**Name** _____

1. Nature of sample (brand, etc.) _____.

    (a) Weight of fresh sample .... _____g.

    (b) Dried weight .... _____g.

    (c) Weight of moisture in sample
    (subtract line 1(b) from line 1(a)) = ... _____g.

    (d) Percent moisture in sample
    (divide line 1(c) by line 1(a) multiply by 100) = ... _____%.

2. "Bread roasting"

    (a) Observations_____.

    (b) Volume of filtrate .... _____mL.

3. Precipitation

    (a) Observations upon adding silver nitrate_____.

    (b) Was any further cloudiness seen when a few drops more of $AgNO_3$ were
    added to the supernatant?_____.

4. Filtration and collection

    (a) Dried weight of paper + silver chloride precipitate .... _____g AgCl
        +paper.

    (b) Dried weight of an empty filter paper .... _____g paper.

    (c) Weight of AgCl in precipitate
    (subtract line 4(b) from line 4(a)) .... _____g AgCl.

    (d) What color changes do you observe in the precipitate?_____.

5. Calculations

(a) Calculate the amount of NaCl necessary to give
your reported weight of AgCl (multiply line 4(c) by 0.41) .... _____g NaCl.

(b) Total amount of NaCl in your original volume of 50 mL
(divide 50 mL by line 2(b) & multiply by line 5(a)) .... _____g NaCl.

(c) Percent salt (NaCl) in original sample
(divide line 5(b) by line 1(a) and multiply by 100) .... _____% NaCl.

(d) % salt in sample stated on package food label (if given) .... _____% NaCl.

6. Conclusions and comments on experiment.

The Staff of Life

Date _____   **Section number** _____   **Name** _____

1. Explain the color change in the AgCl precipitate. (HINT: Reread the background discussion at the beginning of this experiment.)

2. What is the difference between a chlorine atom and a chlorine ion?

3. From what naturally occurring mineral do we obtain practically all of our chlorine?

4. List two nonfood consumer products which contain chlorine (look for a <u>chlor</u> somewhere on the label).

Think, Speculate, Reflect and Ponder

5. Apart from salt and the need for it by most living organisms, what use does chlorine have which you regard as:

   (a) Most beneficial/essential for us and our environment?

   (b) Least beneficial/essential for us and our environment?

6. Why is the ocean salty and where does this salt come from?

7. Why do you think Larry Hamill, the artist, placed a large Cl symbol in the sketch appearing earlier in this experiment?

8. Is using Cl_2 instead of just Cl necessary in the **Larry Hamill** sketch? Explain.

Experiment 12

Meat Analysis: Fat, Water & Protein/Carbohydrate Content (Warning: This experiment may turn you into a vegetarian.)

Samples From Home

Bring EITHER 10 grams of ground meat (for instance hamburger or sausage) OR any other meat/meat substitute which has been finely diced/chopped before coming to lab (wieners, canned ham, bacon, pastrami, veggieburger, etc.). Sample should be capped in a small jar or tightly wrapped in plastic to prevent water evaporation prior to analysis.

Objectives

The techniques of *extraction* and *azeotropic distillation* as defatting and drying methods will be illustrated through their application to the *quantitative* determination of the fat, water and protein/carbohydrate content of a student supplied food product. The results will be used to check both quantitative recovery techniques and, in the case of ground beef and sausage products, compliance of the food with legal regulations governing content. By analyzing different brands, a group of students can determine which brand of a particular product is the best buy based on protein/carbohydrate content.

Background

Since meat and meat products are normally sold by the pound (and maybe by the kilogram also someday soon), most consumers would like to avoid having to pay for excess fat or water—especially inasmuch as these ingredients cost just as much as protein! Sellers, on the other hand, would obviously like to sell as much water and fat as possible at protein prices. In the absence of any regulation, such protein prices for excessive fat and water is just what consumers would likely be gouged for.

The fat and water content of whole meat like steaks and chops are, of course, difficult to alter. Although water can be injected into some whole meat products like hams, altering must occur mainly through the diet and inactivity on the feed lot where most cattle are fattened up before slaughter. In the case of ground and processed meats like hamburger, wieners, sausage, sandwich cold cuts and spreads, however, the sky is the limit when it comes to adding cheap bulking/extender agents, fat and water. Fortunately, because of regulatory agencies, many of these food products must now adhere to legal limits and not "sky limits."

Federal standards require that any product labeled ground beef or hamburger contain not more than 30% total fat. No added water, binders, or extenders (like corn syrup, soybean meal, etc.) may be present. Since meats after grinding do not generally leave state boundaries, state and county agencies may also establish and monitor other "reasonable" standards. Some meat quality designations and fat limits can be seen in Table 12.1. Protein levels in these products will run about $1/4$ of the water content. (For a report on McDonald's McLean Deluxe Burger and similar competitive fast foods, see *Consumer Reports* July, 1991; an analysis of the great American hot dog appears in *Consumer Reports*, July 1993.)

Ground beef and Chuck Label	Maximum % Fat	Typical % Water	Regulatory Agency
Regular	30	55–60	U.S. Department of Agriculture
Lean	23	60–69	Some States and Counties
Extra Lean	16	70	Some States and Counties
Leanest	9		Some States and Counties

Table 12.1. Some Whole Ground Beef Standards (October, 2002).

Federal regulations have been established for many other kinds of meat products as well. Whole ground sausage, for example, must contain only muscle meat and no organs. Some standards for sausage appear in Table 12.2.

Pork sausage, for example, can be expected to contain typically at least 38% fat, which should be no surprise to anyone who has watched the sausage "shrink and swim" during cooking. Cooked pork sausages have been found to average 60% of original raw weight, although results varied widely (*Consumer Reports*, August 1968). For comparison, cooked weight of sliced bacon ranged from 30 to 40% of raw weight (*Consumer Reports*, October, 1989).

Whole Sausage	Maximum % Fat	Comments
Beef	50	Up to 3% Added Water Permitted
Pork	50	Up to 3% Added Water Permitted
Breakfast Sausage	50	Up to 3% Added Water and 3 1/2% extender O.K.

Table 12.2. Some Whole Sausage Regulations (October, 2002).

Another widely sold type of meat product is cooked sausage like hotdogs and liverwurst (see Table 12.3). These come under United States Department of Agriculture (USDA) control as they normally are shipped across state lines. Cooked sausage products may contain no more than 30% fat and, since 1988, have also been subject to the 40% rule. This rule states that the percentage of fat plus the percentage of water added to the product by the processor must total no more than 40%. We can see from this formula that if the processors wish to make you pay for more *added* water (at "meat" prices), they will have to reduce the fat content accordingly. Federal regulations permit these cooked sausage products to contain up to 2 or 3 1/2 % of certain non-meat additives like soybean meal and nonfat dry milk solids. Thus, some of the protein listed on the label may come from these additives instead of from the meat itself.

Cooked Sausage Product	Maximum % Fat	Typical % Water
Hotdog, Vienna, Bologna, Liverwurst, Thuringer	30	55–60
Salami, Cotto Salami, Pepperoni	30	Low

Table 12.3. Some Cooked Sausage Regulations (October, 2002).

Actually, the various regulatory agencies have files of considerable analytical data for a wide variety of food products by brand name. But because these agencies treat the information as highly confidential, it verges on the impossible for the public to obtain access to these data, even though it was acquired through the use of public funds. It appears that a combination of a need to know, litigation, and application through the Freedom of Information Act may be necessary to make this information public. Even these long and very time-consuming processes do not guarantee success.

The background discussion of distillation in Experiment 9: *Alcohol Content of Beverages and Consumer Products* indicates that this technique has indeed many not so simple ramifications. This experiment explores the practical aspects of the boiling together of two *volatile* liquids which, unlike those in the alcohol experiment, are *not* soluble (chemists say *immiscible*) in each other. This process is known as an *azeotropic distillation*. The word azeotropic refers to a constant boiling mixture, while the word distillation, of course, applies to a liquid boiling process.

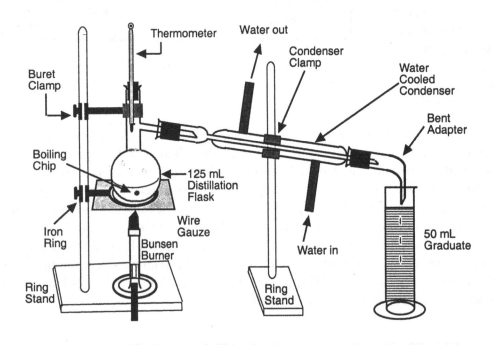

Figure 12.1. A distillation setup.

Applied to your food sample, *azeotropic distillation* affords a neat method of separating the water in a way to permit its rapid *quantitative* measurement, while at the same time also extracting out the fat and leaving behind a cooked, defatted, dehydrated residue of pure protein/carbohydrate. Tetrachloroethylene (a commercial cleaning and degreasing solvent) is a liquid which, when evaporated or boiled, will carry over with its vapors any water which is present. Because tetrachloroethylene is *immiscible* with water, the condensed vapors which drop out of the end of your condenser (the *distillate*) will be cloudy as long as any water is present. As the liquid collects in your receiver, the dispersed micro-droplets of water will slowly come together to form a continuous water layer floating on top of the tetrachloroethylene.

And as an added plus, because tetrachloroethylene is a good fat solvent, the *nonvolatile* fat will dissolve out of the meat into the liquid tetrachloroethylene while in your distillation flask (the fat becomes *extracted* from the meat). When the distillation is stopped, all of the water will have distilled out, while all the fat will remain dissolved in the solution left in your flask. This solution can then be filtered and evaporated to yield the pure fat itself.

In keeping with the importance and theme of recycling our materials as much as possible on spaceship Earth, this experiment is designed so that most of the tetrachloroethylene used is recovered for use again by someone else. This organic solvent is nonflammable, but you should avoid breathing it or unnecessarily exposing its vapors to the air in the lab. Although tetrachloroethylene is less than one-tenth as toxic as the better known carbon tetrachloride, all chlorinated hydrocarbons can cause liver damage upon prolonged breathing and have also been implicated in the destruction of the ozone layer.

Procedure

(See the discussion of distillation in Experiment 9: *Alcohol Content of Beverages and Consumer Products.*)

1. Your sample must be ground, minced, diced, finely chopped or otherwise rendered into particles as small as possible. If you are analyzing a nonuniform product (e.g., a meat like bacon) try to take your sample so that it is as representative as possible of the entire large piece.

2. If not already assembled for you, set up a distillation apparatus as shown on these two pages. Your whole apparatus should be clean, but must be especially bone dry. Acetone may be used **(Caution: Flammable)** to wash any water out of the inside of your glassware and speed drying (check with your lab instructor for help here).

Weigh out about 10 grams of your sample on a square of waxed paper or plastic film to the nearest 0.01 g and place into the bottom of a 125 mL distillation flask. Insert a long stem funnel into the top of your distillation flask and add 50 mL of tetrachloroethylene (TCE) to the sample in the flask using a 50 mL graduated cylinder. The laboratory HOOD should be ON to exhaust any tetrachloroethylene vapors that escape into the air. Use a spatula or solid glass rod if necessary to pull apart your sample if the pieces stick together in a clump. Reassemble your apparatus.

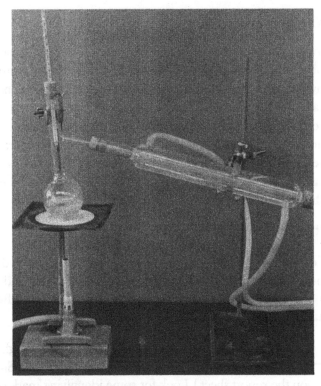

Figure 12.2. Apparatus for an azeotropic distillation.

HAVE THE LAB INSTRUCTOR CHECK YOUR SETUP BEFORE PROCEEDING.

3. Turn on the condenser water until a gentle stream comes out of the exit tube. Bring the solution up to the boiling point quickly using a hot flame. When actual distillation of vapors over into your condenser begins (as seen by a ring of hot vapor rising up the neck of the distillation flask), cut back the heat by partly closing the gas control valve to the Bunsen burner. Distill slowly so that you collect an average of about one drop of distillate every two to three seconds (1 to 1 ¹/₂ mL per minute).

Another way to help control the heat, and hence the rate of distillation, is to slide the wire gauze in and out so as to place open screen (more heat) or white ceramic center (less heat) between the burner flame and distillation flask. By sliding the gauze around you should be able to find just the right position to give you the desired one drop every 2–3 seconds distillation rate.

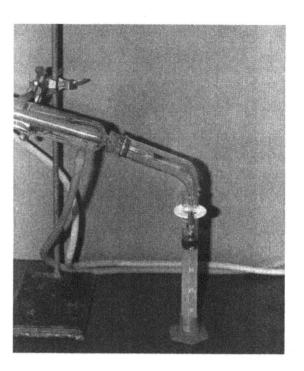

Figure 12.3. Collecting distilled TCE and water with dye added to water layer.

What do you observe?

Continue distilling at this rate until 40 mL of liquid tetrachloroethylene (the bottom layer) has been collected in your graduate. This should take 30–45 minutes, by which time your distillate should have become almost, if not completely, clear, and you have a distinct top layer of water in your graduate. If you are using a thermometer, the initial vapor temperature will hold around 87–95 °C, creeping up towards 110 °C after 20–30 mL of distillate have been collected. Most of the water should be removed by then, but try to keep the temperature below 110 °C as long as possible. By the time your final total lower layer volume of 40 mL is reached, the temperature should have reached at least 118 °C or higher if all the water has distilled over as it should have (tetrachloroethylene itself boils at 121 °C).

STOP the distillation at this point.

It cannot be overemphasized that if you try to rush the distillation, you will not remove all the water and your distillate (the liquid dropping out of your condenser) will still be cloudy at the end of the distillation. The accuracy of your results would then suffer. Even in chemistry—especially so—haste makes waste!

While the distillation is underway, clean, dry and weigh an evaporating dish to the nearest 0.01 g so it will be ready for the *Fat Content Determination* in Step 5. (You can enter this weight on line 5(a) on the report sheet.) Look for some identifying mark or number on the balance so you can use the same one when you reweigh this dish later.

Water Content Determination

After completing your distillation, carefully disassemble the apparatus so you can tip up the condenser and permit any water droplets "hung up" in it to run out into your graduate. SAVE the contents in your distillation flask for the *Fat Content and Protein/Carbohydrate Content Determinations*.

If you (or your lab instructor) have not already done so during the distillation, add a micro-pinch (about $^1/_4$ the size of a BB) of a water soluble dye to the liquid in your graduate. This will dissolve only in the water layer and make it easier to see during the separation procedure which follows. If necessary, use a thin solid glass rod to reach down into the liquid and dislodge any obvious droplets of water which remain stuck to the sides of the graduate down in the tetrachloroethylene layer.

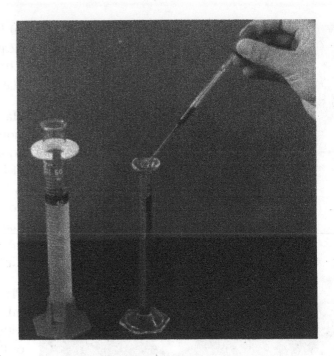

Figure 12.4. Transferring top water layer to a small graduate.

4. Using a long-nose medicine dropper, transfer all of the now colored top layer in your 50 mL graduate to a dry 10 mL graduate that has 0.2 mL graduations.

   (a) Read the total volume of your colored water to the nearest 0.1 mL and record on the report sheet. (Remember to read the *bottom* of the meniscus as shown in Experiment 2: *Going Metric*.) Discard the water layer and pour the 40 mL of TCE left in your graduate into the "Waste TCE" container provided in the lab.

Fat Content Determination

Now back to the contents of your distillation flask. If you have measured all of your initial volumes correctly, you should have about 10 mL of tetrachloroethylene containing the fat in your distillation flask, along with insoluble chunks of defatted, dehydrated, cooked protein/carbohydrate residue.

5. Gravity filter the distillation flask contents (solid pieces and liquid) by pouring them into a small filter paper cone (Whatman #1, 9 cm) mounted in your funnel held in an iron ring. (Refer to the experimental procedure in Experiment 11: *The Staff of Life and Chlorine* for the gravity filtration procedure). Do not try to "wet" your filter paper with water; the liquid being filtered in this case is not water!

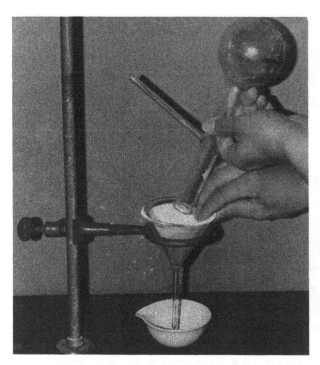

Figure 12.5. Filtering fat containing tetrachloroethylene liquid from solid protein/carbohydrate.

(a) If you have not already done so, weigh a clean, dry evaporating dish to the nearest 0.01 g and enter data onto report sheet line 5(a).

Have the filtrate drop directly into your previously weighed evaporating dish. After the liquid has all run through, add about 3 mL of fresh tetrachloroethylene to the flask, swirl to rinse the sides and any residue, and pour into the filter paper cone in a manner that also washes the sides of the paper. SAVE all solid residue pieces for the *Protein/Carbohydrate Content Determination* in Step 6.

Place the evaporating dish containing the 13 mL of fat solution onto a piece of paper with your name on it and set it in the back of the hood until the next laboratory period. By then the tetrachloroethylene solvent will have evaporated leaving the pure fat ready to weigh and examine.

Figure 12.6. Fat residue ready for weighing.

Note: Steps 5(b) and 5(c) will be done at the beginning of the NEXT laboratory period.

5 (b). NEXT LAB PERIOD—Weigh evaporating dish containing the fat.

5 (c). NEXT LAB PERIOD—Note nature of the fat residue.

Protein/Carbohydrate Content Determination

6. Transfer all the chunks of your insoluble protein/carbohydrate residue left from your filtration into a 50 mL beaker. Place this beaker in the hood alongside your evaporating dish to air dry until the next laboratory period.

(a) NEXT LAB PERIOD—Weigh your hard protein/carbohydrate chunks and report the yield.

(b) NEXT LAB PERIOD—Note the nature and appearance of your residue. You have prepared a cooked, defatted, and dehydrated food product.

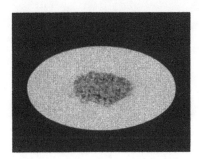

Figure 12.7. Protein/carbo-
hydrate chunks.

NOTE: The calculations called for on the report sheet will, in addition to assessing ingredient amounts (items number 7 and 8 in the calculations), also permit you to evaluate your own ability as a *quantitative* chemist (item number 10 in the calculations). Your total weight of water + fat + protein/carbohydrate would equal your 10 grams of original sample if you achieved 100% recovery. Although it is possible to get good recoveries of fat and solids, expect about 10% of the water content to be lost during isolation (primarily through wetting of the condenser and upper part of the distillation flask). Thus your percent water in line 8a on the calculation page will represent only about 90% of original water in the meat sample. A closer estimate of the total water present can consequently be obtained by multiplying your per cent water by 1.1 as is done in the calculation on line 8(a).

Meat Analysis

Date _____   **Section number** _____   **Name** _____

DATA

1. Sample Information

    (a) Kind of Sample_____

       Brand_____

       Grade (Regular, Lean, etc.)_____

       Source of Sample (Where Purchased)_____

    (b) Price of sample package or can_____

    (c) Weight of sample package or can_____

2. Weight of sample taken for analysis (to nearest 0.01 g) .... _____g.

3. Comments on initial appearance of distillate _____

4. *Water Content Determination*

    (a) Accurately measured volume of water layer
    (to nearest 0.1 mL) .... _____mL.

5. *Fat Content Determination*

    (a) Weight of empty evaporating dish (to nearest 0.01 g) .... _____g.

    (b) Weight of evaporating dish and fat residue
    (to nearest 0.01 g) .... _____g.

    (c) Nature, appearance, smell of fatty residue.

    _____

6. *Protein/Carbohydrate Determination*

    (a) Weight of protein/carbohydrate residue .... _____g.

    (b) Nature and appearance of solid residue

    _____

CALCULATIONS

7.   (a) Weight of water
(line 4(a) multiplied by the density of water, 1 g/mL) = ... _____g H_2O.

   (b) Weight of fat (line 5(a) subtracted from line 5(b)) .... _____g fat.

   (c) Weight of protein/carbohydrate (line 6(a)) .... _____g prot/carb.

8.   (a) Percent water in sample
(divide line 7(a) by line 2 and multiply by 110) = ... _____% water.

   (b) Percent fat in sample
(divide line 7(b) by line 2 and multiply by 100) = ... _____% fat.

   (c) Percent protein/carbohydrate in sample
(divide line 7(c) by line 2 and multiply by 100) = ... _____% prot/carb.

   (d) Are these values within legal/typical limits set by regulatory agencies?
(See tables in the background to this experiment.)

---

9.   (a) Price per pound of purchased sample—look on the
package label or calculate: (divide line 1(b) by line 1(c)) = ... _____cents/pound.

   (b) Price per pound of just the pure protein/carbohydrate in your sample
(divide line 9(a) by line 8(c) and multiply by 100 = ... _____cents/pound.

   (c) How do your results in 9(b) compare with those for brands of similar
products examined by other students, if any?

10.   (a) Total *quantitative* recovery (add lines 7a + 7b + 7c) = ... _____g.

   (b) Percent recovery
(divide line 10(a) by line 2 and multiply by 100) = ... _____%.

   (c) What do the results in 10(b) tell you about your *quantitative* technique?
(See "NOTE" at the end of the procedure section.)

---

11.   Comments and conclusions on experiment

Meat Analysis

Date _____   **Section number** _____   **Name** _____

1. From your observations during this experiment, does tetrachloroethylene or water have the higher density? Justify your choice.

2. Describe what substance makes your tetrachloroethylene distillate cloudy, and explain why this substance causes such cloudiness. (Hint: See the background discussion to this experiment.)

3. What new regulations for the food product that you analyzed in this experiment would you like to see enacted, if any?

Think, Speculate, Reflect and Ponder

4. Protein is found in foods other than meat.

      (a) What foods are rich in protein besides meat products?

      (b) Why are non-meat protein foods a much more efficient protein source for man than meat? (Hint. Consider the energy requirements and land & water needed for growing.)

      (c) In spite of the efficiencies mentioned in part (b), consumers in the United States are still basically meat eaters. Why?

5. What will be the effect on the planet if all of the underdeveloped countries adopt the same meat production methods (energy, land and water resource requirements) that the United States uses now?

Experiment 13

Vitamin C in Your Diet?

Sample From Home

Bring a liquid juice containing vitamin C. Citrus juices such as orange and grapefruit—either fresh, frozen, canned, or powder—usually contain large amounts of vitamin C and work well in this experiment. Lime and lemon juices, on the other hand, often contain small amounts; apple, apricot, papaya, lime and white grape juices usually have little if any vitamin C content unless added artificially by the bottler. About 250 mL ($^1/_2$ pint) of "drinking concentration" juice is desirable; half this volume would be an absolute minimum.

NOTE. It is important that the color of the juice not interfere with the detection of the pink color formed in the reaction. Red or intensely colored juices must be avoided. Tomato type or purple grape juices, for example, would be obvious no-no's.

Objectives

An official governmental procedure will be used to determine *quantitatively* the vitamin C content of a sample from home. This procedure will illustrate the analytical technique of *titration* using color indicators. Calculations based upon the titration results will permit a judgment to be made regarding the importance of the particular food sample examined as a source of vitamin C. The results can also be compared to the manufacturer's claimed vitamin C content when given on the label.

Background

If one were to heed the rhetoric of the drug and pharmaceutical industry, the only sure way to avoid malnutrition and disease is to become part of a 100% pill popping society. The *organic* drug manufacturers take things one step further by saying that the only good food, vitamin or drug is an *organic* one—not something made from foreign chemicals or dirty old gooey crude oil! We are socked with advertisements even for organic cosmetics and hair shampoos. And whole *organic* vitamin display sections now can be found in many large stores, where rose hip vitamin C and sea salt vie with their supposed *nonorganic* or unnatural counterparts for the consumer's dollar—and a lot more of it!

Vitamins are certainly important to individuals who suffer from malnutrition or specific biochemical disorders. These persons represent, however, only a tiny fraction of those who regularly take such dietary supplements. For the great masses, it cannot be denied that vitamins can be effective as placebos—for instance a psychological uplift to the person who *believes* that they will help. In spite of the billions of dollars spent annually on vitamins and mineral supplements, vitamin research indicates that most persons making these purchases are simply padding the financial pockets of the drug manufacturers for such psychological uplift. It has furthermore been shown that there can be too much of a good thing even with the vitamin panacea. Evidence has been published that points to the danger of ingesting too much of particular vitamins (specifically vitamins A and D) by the overuse of vitamin supplements. (See *Consumer Reports*, September, 1994, for a discussion of vitamins.)

> ## Nothing Exists Except Atoms and Empty Space
> ### (all else is opinion)
> Democritus 460–370 B.C.

As a type of compound, vitamins differ widely in their chemical structures. Physiologically, they may be defined as organic compounds that, while essential constituents of the diet, are required in only minute amounts. They thus differ from hormones in that the body cannot synthesize them and they differ from trace elements because they are organic (carbon compounds). They also differ from fats, proteins and carbohydrates because they are required in very small amounts. In an average daily diet of 600 g (on a dry basis), the total vitamin intake would represent only 0.1 to 0.2 grams. It is because vitamins play an essentially *catalytic* role in life processes that they can be effective in such small doses.

Although claims for virility and sexual potency have not been touted (yet?) for vitamin C, this vitamin has received considerable attention in connection with general resistance to disease and particularly that elusive, incurable scourge—the common cold. No less a luminary than Dr. Linus Pauling (1901–1994) maintained for many years that daily doses of several thousand milligrams of vitamin C could prevent the common cold. (Regular vitamin C supplements include only a few hundred milligrams.) Linus, himself, took up to 20,000 mg.

Dr. Pauling was awarded Nobel prizes in both bio-chemistry and peace - the only person ever to receive two unshared prizes. And had Dr. Pauling not been blacklisted as a communist sympathizer, he might well have at least shared in a third Nobel prize for discovering the structure of the DNA double helix (heredity secret). But the special X-ray pictures upon which the discovery was based were in England and he was denied a travel visa.

LINUS PAULING ON HISTORY

A patient tells the doctor that he is taking vitamin C at more than 1000 mg a day on the recommendation of Linus Pauling. The doctor responds:

In 1973, "That quack! It'll kill you."
In 1983, "O.K., it won't hurt."
In 1993, "O.K, it might help."
In 2003, " ? ? ? "

To date, the best evidence seems to indicate that vitamin C might indeed reduce the *symptoms* of colds, but not their *incidence*. But more hard evidence on this controversial question is still needed, for the long-range effects of the relatively massive doses called for by Dr. Pauling are not clear. Most excess vitamin C seems to be simply excreted unchanged in the urine. In the pure state, it is a colorless, crystalline solid.

Part of the problem is due to the practical difficulty of controlling and obtaining meaningful test results "in the field" (that is, out in the population). Compounding this difficulty is the fact that, although important in promoting healing and fighting infection, the actual biochemical role of vitamin C (or L-ascorbic acid as it is chemically known) is still far from clear. Historically, we do know that the absence of adequate amounts of ascorbic acid in the diet results in the dietary disease called scurvy. The sailors who spent many months at sea on early sailing ships much feared this disease until it was discovered that including fruit in the diet effectively prevented scurvy. The British Navy's solution to this was to include vitamin C rich limes in the food stores shipped on long voyages. The widespread eating of limes by the British Navy's seamen earned them the nickname "limeys."

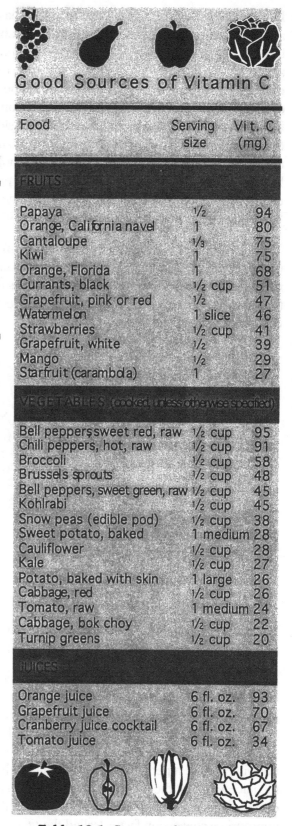

Good Sources of Vitamin C

Food	Serving size	Vit. C (mg)
FRUITS		
Papaya	½	94
Orange, California navel	1	80
Cantaloupe	⅓	75
Kiwi	1	75
Orange, Florida	1	68
Currants, black	½ cup	51
Grapefruit, pink or red	½	47
Watermelon	1 slice	46
Strawberries	½ cup	41
Grapefruit, white	½	39
Mango	½	29
Starfruit (carambola)	1	27
VEGETABLES (cooked unless otherwise specified)		
Bell peppers, sweet red, raw	½ cup	95
Chili peppers, hot, raw	½ cup	91
Broccoli	½ cup	58
Brussels sprouts	½ cup	48
Bell peppers, sweet green, raw	½ cup	45
Kohlrabi	½ cup	45
Snow peas (edible pod)	½ cup	38
Sweet potato, baked	1 medium	28
Cauliflower	½ cup	28
Kale	½ cup	27
Potato, baked with skin	1 large	26
Cabbage, red	½ cup	26
Tomato, raw	1 medium	24
Cabbage, bok choy	½ cup	22
Turnip greens	½ cup	20
JUICES		
Orange juice	6 fl. oz.	93
Grapefruit juice	6 fl. oz.	70
Cranberry juice cocktail	6 fl. oz.	67
Tomato juice	6 fl. oz.	34

Table 13.1. Sources of Vitamin C.

Paprika (obtained from dried sweet peppers) has perhaps the highest concentration of vitamin C, although as a class of foods, citrus fruits are probably most important. And because of their prevalence in the American diet, potatoes as well represent a significant source for this vitamin. Ascorbic acid is known chemically as a good *reducing* agent, which means that it undergoes *oxidation* very easily, even reacting with oxygen in the air. (The terms *oxidation* and *reduction* refer to the tendency of a chemical to *give up* or *attract* electrons.) This is why most of the vitamin C content of foods is lost during cooking. This can happen slowly when ascorbic acid is exposed to oxygen in the air, and is accelerated greatly by heating in air, or by other chemical oxidizing agents. Most of the vitamin C content of foods is lost during cooking. This special reactivity (oxidizability) of ascorbic acid is made use of in this experiment: the particular oxidizing agent (indophenol) that we will use was chosen in part because the color changes accompanying the reaction make it possible to "see" how much vitamin C is present in our test sample.

The method used for your ascorbic acid determination is taken directly from the official procedure used by the U.S. Food and Drug Administration (FDA) and other regulatory agencies. During the reaction, the intense blue indophenol becomes decolorized as it is added to the ascorbic acid solution. So that you have the information to impress all your friends and those around you, here is what the chemical picture looks like:

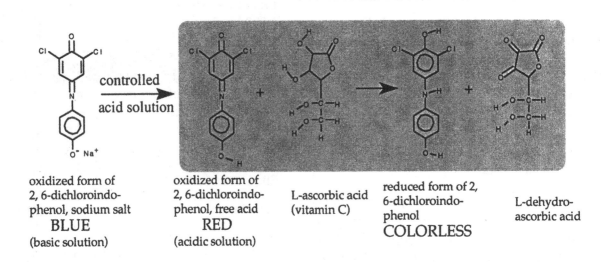

oxidized form of 2, 6-dichloroindophenol, sodium salt
BLUE
(basic solution)

oxidized form of 2, 6-dichloroindophenol, free acid
RED
(acidic solution)

L-ascorbic acid (vitamin C)

reduced form of 2, 6-dichloroindophenol
COLORLESS

L-dehydroascorbic acid

The shaded box represents the reaction of indophenol with ascorbic acid. (Each corner of the polygons represents a carbon atom unless otherwise indicated.)

As the indophenol *blue* is added to the acidified juice sample, it would form a visible indophenol *red* color if it were not immediately oxidized to the indophenol *colorless* form by the ascorbic acid present. (As you might guess by looking at the equations, the color and light absorbing properties of molecules can be dramatically altered by subtle changes in their structure.) At the exact point when the last bit of ascorbic acid has reacted, further addition of just a drop or two of the indophenol will produce a persisting *red* color due to the indophenol red form. In the controlled acidity of the juice solution, the indophenol turns *red* much like blue litmus paper turns red in acid. The indophenol is *blue* in your burette only because that solution is slightly *basic* (alkaline). Using an appropriate conversion factor (called the *titer*), this *quantitative* amount of indophenol used in the reaction can be converted into the amount of ascorbic acid (vitamin C) present in the sample.

It is desirable to use large volumes of juice in the experiment in order to reduce oxidation (and hence the destruction) of the ascorbic acid by the oxygen in the air prior to analysis. This method of analysis gives reliable results provided that the juice does not contain materials which (like ascorbic acid) also are reducing agents. Such examples include oxidizable forms of iron, tin, and copper; sulfur dioxide and sulfite ions also interfere with an accurate determination.

Procedure

1. When using frozen juices or drink powders, you should add enough water to dilute it up to "drinking strength" before proceeding. State the nature of your sample on the report sheet. (If your juice is not free of sediment, the pulpy juice should first be strained through several layers of cheesecloth held in place with a rubber band over the mouth of a 250 or 400 mL beaker.)

Using a 50 mL graduated cylinder, measure 50 mL of your clear or strained juice into a 250 mL Erlenmeyer flask; using the same graduate, add 50 mL of the metaphosphoric acid/acetic acid reagent   to the Erlenmeyer and swirl to mix well.

Figure 13.1. Straining a pulpy juice through cheesecloth.

Obtain a 25 or 50 mL burette and a double burette clamp and set up for a titration at your work
bench. Your lab instructor will illustrate the techniques of this procedure. Basically,
the burette is just a tall, thin graduated cylinder with a stopcock (valve)
at the bottom which can measure out
liquids (through this stopcock) to a high degree of
accuracy—commonly to the nearest 0.01 mL. The
process of *quantitatively* reacting or neutralizing
another substance by metering out the exact required amount of liquid with a burette is called a
titration.

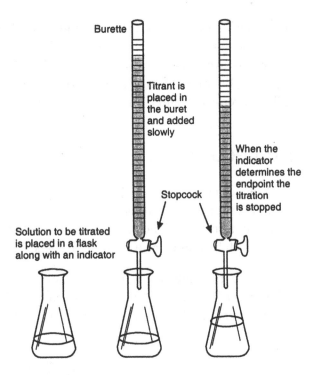

Figure 13.2. The titration process.

Check your stopcock to see that it turns freely.
Ground glass stopcocks (but not those made of
Teflon® plastic) may require a little grease, but be
careful. Too much grease will easily plug up the
burette tip; see your lab instructor for help, if
necessary. If your burette is clean and
dry, consider yourself lucky. If not,
wash the burette with soap and water, rinse with a little distilled water and permit it to
drain. A clean burette will not leave drops of water clinging to the sides. Lastly, rinse the
burette with a few milliliters of the deep blue **indophenol reagent** to wash away any water left in
your burette.

Clamp the burette in its holder as shown in Figure 13.2 and fill it with **indophenol reagent** to
about an inch above the 0.00 mL mark. Place a waste beaker under the burette and open the stopcock and allow this liquid to flush the burette tip and wash out any air bubbles. When the meniscus level (recall Experiment 2: *Going Metric with the Rest of the World*) drops down even with the
0.00 mL mark, shut off the stopcock and discard the liquid that has drained out into the appropriate waste container. Examine the graduations on your burette carefully—both their numerical equivalent and the direction in which the numbers are increasing—to be sure you
will read the meniscus levels correctly during the rest of the experiment. Ask your lab instructor if you have any questions or doubts on this before proceeding.

Using a 10 mL pipette with a pipetter bulb attached (a type like the Propipetter® works well), transfer 10.0 mL of your juice + **metaphosphoric acid/acetic acid** mixture to a 50 mL Erlenmeyer flask. (Your lab instructor can demonstrate.) Place this flask onto a piece of white paper underneath your burette and proceed to *titrate* your juice + acid sample with the intense
blue **indophenol** as rapidly as possible until a light but distinct rose-pink color persists
for at least 5 seconds. The flask contents must also be continuously swirled to create
proper mixing throughout the titration. The point at which a rose-pink color persists for
at least 5 seconds is called the *endpoint* of your titration.

With most samples, the titration is begun by fully opening the burette stopcock. You will see that the solution in your flask immediately surrounding the stream of added indophenol turns pink for just a moment. With a little practice you will recognize how the slowness with which this pink color disappears can indicate how close you are to the final endpoint.

As practice will show, you need to slow down as you approach the endpoint to a drop-by-drop addition. Remember: the flask contents must be continuously swirled to create proper mixing throughout the titration. Note the final burette reading and, after subtracting the initial reading, record the total net volume of indophenol reagent added. You can expect 10–20 mL of indophenol reagent to be required per each 10 mL sample (called an *aliquot*). Record all volumes on the report sheet to the nearest 0.1 mL. (If you "overshoot" the endpoint, you will have to repeat the titration as described in the next step 2.)

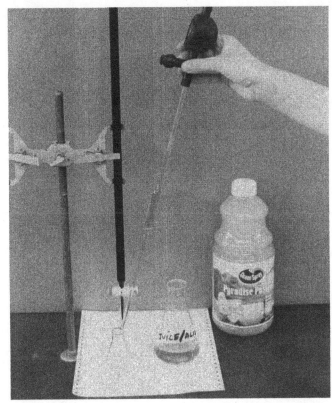

Figure 13.3. Pipetting juice + acid mixture into flask.

2. At the completion of a titration, discard the contents of the Erlenmeyer flask in an appropriate waste container and rinse the flask thoroughly with distilled water. After doing this, repeat the previous procedure by titrating similar 10 mL aliquots of your juice + acid sample until you obtain two titration volumes which agree within 0.5 mL of each other. Record all your data on the report sheet. Note that in successive titrations, the *final* burette reading from the previous run becomes the *initial* reading in the next run. This means that you don't have to fill up your burette to the 0.00 mL mark before each titration unless you will be exceeding the volume of reagent left in the burette during the next trial. You can now appreciate that the more highly colored your original solution is, the more difficult it will be to see a sharp color change at the endpoint.

3. Your lab instructor will give you the conversion factor relating the volume of indophenol reagent to the milligrams of vitamin C present (the *titer* of the indophenol reagent). Use this *titer* factor to perform the calculations called for on the report sheet. From the results, state your conclusions regarding the importance of a glass of your juice as a vitamin C source in your diet. Where possible, compare your results with those stated on the product label.

Vitamin C in Your Diet?

Date _____   **Section number** _____   **Name** _____

1. Nature of Product

        Appearance _____.

        Brand _____.

        Source of Sample _____.

2. Data: (Record all volumes to nearest 0.1 mL)

    (a) Volumes of indophenol reagent required
    to titrate 10 mL samples of juice

	Trial 1	Trial 2	Trial 3 (if needed)
(Final burette reading)	_____mL	_____mL	_____mL.
(Initial burette reading)	_____mL	_____mL	_____mL.
(Net volume indophenol used)	_____mL	_____mL	_____mL.

(b) Average volume of indophenol used ...._____mL.

(c) "Blank" (Volume of indophenol necessary
to give pink endpoint if no vitamin C were present) ....___0.10___ mL.

(d) Net volume of indophenol required to titrate
only the vitamin C present in sample
(subtract line 2c from line 2b) = ..._____mL.

3. Calculations:

(a) Titer of indophenol reagent
(mg of vitamin C equivalent to each 1.00 mL of the
reagent—get this from your lab instructor) .... _____mg/mL.

(b) Milligrams of vitamin C present in 5 mL of juice
present in the 10 mL of juice/acid mixture
(multiply line 2d by line 3a) = ... _____mg vit. C.
(per 5 mL of juice)

(c) Amount of vitamin C in an average
serving (8 oz glassful = 240 mL)
(multiply 240 mL times line 3(b)
and then divide by 5) = ... _____mg vit. C.
(per 8 oz glassful).

(d) How many 8 oz. glasses of juice would be
required to obtain 100% of the recommended
daily adult allowance (RDA) of 60 mg
(divide 60 mg by line 3(c)) = ... _____glassfuls.

(e) What percent of the RDA would be obtained
by drinking just one 8 oz glassful?
(divide line 3(c) by 60 & multiply by 100) .... _____%.

(f) What is the claimed % RDA for vitamin C
from a 8 oz serving appearing on the label, if any .... _____%.

Note: if the serving size stated on the label is not 8 oz, the answer to 3(f) will have to be calculated. Multiply the %RDA on label by 8 and then divide by the serving size in oz. For example, if the stated RDA is 80% for a 6 oz serving, the %RDA for an 8 oz glassful would be

$$\frac{80\% \times 8}{6} = 107\%$$

4. Comments and conclusions on experiment

Vitamin C in Your Diet?

Date _____   **Section number** _____   **Name** _____

1. A precaution commonly found on bottles of ascorbic acid tablets says to keep the bottle tightly capped. Furthermore, the bottle itself is almost always dark brown or amber in color. Why?

2. One possible biochemical role sometimes suggested for ascorbic acid is its ability to prevent, or at least diminish, the oxidation of other critical substances crucial to normal body function; this role thereby acts to protect them from oxidative destruction. The ascorbic acid would thus be functioning as an "*anti-oxidant*," a role some say supports its suggested use as a possible *anti-aging* drug also. Explain the rationale for this. (Hint: a person sacrificing his life for someone else might be chemically analogous to the role of ascorbic acid in our body.)

3. In some geographical areas, persons used to be afflicted with a problem called goiter due to a lack of iodide ion in the food they ate. Nowadays this problem no longer exists.

   (a) Could the lack of iodide in the diet be classified as a vitamin deficiency? Explain.

   (b) Why does this problem no longer exist?

Think, Speculate, Reflect and Ponder

4. The danger of overdosing on vitamins seems to concern primarily just the oil/fat soluble vitamins.

    (a) Which vitamins are these?

    (b) How might the amounts of these vitamins build up in the body?

5. Taking large doses of vitamin C has been known to cause a decrease in the amount of B vitamins in the body. Knowing this, do you think B vitamins are more <u>fat</u> soluble or <u>water</u> soluble? Why?

Experiment 14

Label Reading:
Know What You Buy, Use and Eat

Samples From Home

This is a dry-lab experiment, but you will need to look at the following:

> 1. Products found in your kitchen, garage, medicine cabinet, grocery, discount, hardware or garden store.

> 2. *The Merck Index* which can be found in the laboratory and in many libraries.

Objectives

The importance of reading labels on things we buy will be emphasized, as well as the usefulness of *The Merck Index* as a general source book for looking up the properties of label ingredients. The chemical and physiological properties of some selected drug, food additive, consumer product and "cide" chemicals will be researched as a library assignment.

Background

A well known furniture polish contained chemicals which are classified as dangerous both by breathing and skin contact. A widely advertised insecticide for continuous household use still has as its active ingredient a close nerve gas relative. A very well-known headache formulation some years ago heralded their "new and improved" product to the public, not mentioning that the improvement resulted from the forced removal of one of the drug's ingredients due to its implication in liver disease and cancer from prolonged use.

Figure 14.1. Looking up label ingredients in a reference book.

In all of these cases, the active ingredient in question was listed on the label. Admittedly, chances are that the chemical names will be mostly "Greek" to the average person. Most consumers will simply either not bother to read the label ingredients at all, or stop when reaching a word they don't understand (which may not take very long). After completing this assignment, the chemical names may still look like "Greek" to you, but at least you should know where to go to get some answers about their properties.

All this is not to say that we would, or should, do without many of the ingredients in things we buy and eat. These ingredients can indeed give us "Better things for better living—through chemistry", to use a phrase much heralded in the advertising of the Du Pont Chemical Company before the word *chemical* became *non-grata* in the public eye. Most of us need not, nor would not, want to return to the oft, but ill-named, "good old days" of natural chemicals, foods, and products. Neither would most of us want to live amidst *natural* 14 hour work days with *natural* 45 year life spans, *natural* health care, sickness, pestilence and early death!

Those individuals in the *natural* movement would do well to ask how much more *natural* their rose hip Vitamin C is than Vitamin C derived from glucose; how much better *organic* foods, cosmetics, or hair lotions are than the *nonorganic* variety; how much better Bayer aspirin is than Brand X aspirin? Do these questions deal with fad or fact, truth or fiction? Do the properties of a chemical substance depend upon where its constituent atoms come from? How do "new" atoms differ from "old or used" atoms (recall Experiment 3: *Recycling Aluminum Chemically*)? And what, indeed, is even meant by *natural* or *organic*? Are they just glittering catch words which advertisers hope will empty the pocketbooks of a gullible public?

We frequently read of products touted to "Contain no chemicals!" Really. Ask yourself what this means (you'll have a chance to answer on the question sheet). Chemicals can, of course, be used to adulterate a product (grain, fat, water mixed into ground meat products) and to conceal inferiority or damage in a product (food coloring of meats, fruits, and vegetables). Most alarmingly, chemicals can also be used to promote an immediate "selling" effect to a product without sufficient or responsible evaluation of possible toxic effects on the user (2,2-dichlorovinyl dimethyl phosphate in pest strips or nitrobenzene in furniture

Spicy Spray will make any Attacker see Red

The sprays, manufactured by a Montana company, contain a hot red-pepper concentrate as the active ingredient. Although similar in concept and effect, *Counter Assault* is different from *Mace* because it uses no chemicals and is organic.

List of Import-Prohibited & Restricted Items
Republic of Maldives (1999)

1. Narcotics, illicit drugs and psychotropic substances
2. Pornographic materials
3. Materials contrary to Islam
4. Idols for worship
5. Pork and pork products
6. Alcohol
7. Dogs
8. Dangerous animals
9. Gunpowder and explosives
10. Weapons, firearms and ammunition
11. Spear guns
12. C h e m i c a l s

Table 14.1. Maldivian import restrictions.

polish). Even when precautions *are* written on labels in print large enough to read, too many persons either ignore, or seem contemptuous of, these warnings.

For example, several consumer products labelled "pest strips" give off the nerve gas relative *dichlorovos (DDVP or 2,2-dichlorovinyl dimethyl phosphate).* The original "Shell No-Pest Strips" later morphed into Bio-Strip *"Pest Strip"* and Loveland Industries *"Pest Strip"*. Now, in 2002, similar products include a made in England *"Hot Shot No Pest Strip"* (United Industries) and a USA made clone, *"Revenge Bug Strip"* (Roxide International).

These products are advertised "to kill insects in your home for up to four months". Large bold letters on the front of the package announce that it is for continuous use in homes, cabins, garages, campers, basements, apartments, etc. However, on the back of the package, in small print, reads "Do not use in any room where infants, the sick or aged are, or will be present, for any extended period of confinement". But if you live in your home, you will also be breathing these same toxic vapors. Consumers Union has been warning against these products since 1967 (see *Consumer Reports,* May, 1988 and July, 1990); the state of California has published warnings declaring that "These products (no-pest strips) contain a chemical (DDVP) known to cause cancer and constitute an immediate health hazard".

A more enlightened awareness and concern for chemical safety on the part of industries and the public are resulting in continued changes in the marketplace. Examples of these changes are the three hazardous products mentioned in the opening background paragraph: furniture polish, headache formulation and household insecticide (see Figure 14.2 on next page). Both of the potentially hazardous chemicals formerly present in the widely sold product *Scott's Liquid Gold* furniture polish have now been removed from the formulation. The nitrobenzene disappeared in 1983 and the trichloroethane in 1991. And what about that "new and improved" headache product? *Anacin* used to be just a high-profile brand name for what were generically called APC tablets containing aspirin, phenacetin and caffeine. In this case, less was better. Since becoming "new and improved," *Anacin* has discontinued the use of the suspect ingredient phenacetin. The insecticide pest strips mentioned earlier, however, continue to be readily available.

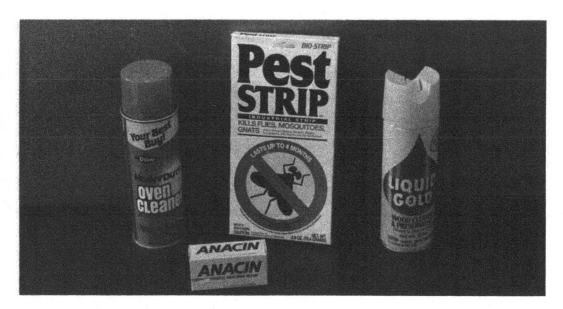

Figure 14.2. Object lessons: Scott's Liquid Gold, Anacin, No-Pest Strip and an oven cleaner.

But perhaps the most potentially hazardous chemical in your entire house (and indeed more dangerous than anything you can see around you or will use in the lab) can likely be found not in the garage or an outbuilding, but right in the kitchen. This chemical "bomb" is the aerosol can of oven cleaner containing sodium hydroxide (regular lye) or potassium hydroxide (super-strong lye). These ingredients are made doubly hazardous by the pressurized container in which they come. Read the label closely; if you think the warning sounds bad, just look up the properties of the active ingredient using the skills that you gain during this lab. Nasty cleaning jobs require nasty chemicals, and sodium or potassium hydroxide does indeed do a good job. But if the can nozzle is depressed while mistakenly aimed at your face, the lye will also do a good job on your eyes and quite possibly damage them for life.

Procedure

 This is really a library assignment. Two different report sheets will be found at the end of this "experiment": 14a and 14b . Your lab instructor will indicate which ONE of these should be completed and turned in, or may just let you choose.

You will need to use *The Merck Index* handbook to complete information called for on the report sheets. Most of this book consists of an alphabetical listing and description of chemicals according to their most common or chemically correct name. If you cannot find what you are looking for in the main listing, check the cross index in the back of the book. Most substances have two or more possible names and this cross index lists many such synonyms with their locator numbers for the main listing where you can find the actual description of the substance. The locator numbers are black bold numbers assigned to each substance entry in *The Merck Index*. Copies of *The Merck Index* can be found in science laboratories, college libraries, many city and county libraries.

THE MERCK INDEX

AN ENCYCLOPEDIA OF CHEMICALS, DRUGS, AND BIOLOGICALS

873. Aspirin. *2-(Acetyloxy)benzoic acid; salicylic acid acetate;* 2-acetoxybenzoic acid; acidum acetylsalicylicum;

acetylsalicylic acid; Acetilum Acidulatum; Acenterine; Aceticyl; Acetophen; Acetosal; Acetosalic Acid; Acetosalin; Acetylin; Acetyl-SAL; Acimetten; Acylpyrin; A.S.A.; Asatard; Aspro; Asteric; Caprin; Claradin; Colfarit; Contrheuma retard; Cosprin; Delgesic; Duramax; ECM; Ecotrin; Empirin; Encaprin; Endydol; Entrophen; Enterosarine; Helicon; Levius; Longasa; Measurin; Neuronika; Platet; Rhodine; Salacetin; Salcetogen; Saletin; Solprin; Solpyron; Xaxa. $C_9H_8O_4$; mol wt 180.15. C 60.00%, H 4.48%, O 35.53%. Prepn: C. Gerhardt, *Ann.* **87**, 149 (1853). Manuf from salicylic acid and acetic anhydride: Faith, Keyes & Clark's *Industrial Chemicals*, F. A. Lowenheim, M. K. Moran, Eds. (Wiley-Interscience, New York, 4th ed., 1975) pp 117-120. Crystallization from acetone: Hamer, Phillips, U.S. pat. **2,890,240** (1959 to Monsanto). Novel process involving distillation: Edmunds, U.S. pat. **3,235,583** (1966 to Norwich Pharm.). Crystal structure: P. J. Wheatley, *J. Chem. Soc. (Suppl.)* **1964**, 6036. Toxicity data: E. R. Hart, *J. Pharmacol. Exp. Ther.* **89**, 205 (1947). Evaluation as a risk factor in Reye's syndrome: P. J. Waldman *et al.*, *J. Am. Med. Assoc.* **247**, 3089 (1982). Review of clinical trials in prevention of myocardial infarction and stroke: P. C. Elwood, *Drugs* **28**, 1-5 (1984). Symposium on aspirin therapy: *Am. J. Med.* **74**, no. 6A, 1-109 (1983). Comprehensive description: K. Florey, Ed. in *Analytical Profiles of Drug Substances*, vol. 8 (Academic Press, New York, 1979) pp 1-46. Monograph: M. J. H. Smith, P. K. Smith, *The Salicylates* (Interscience, New York, 1966) 313 pp. Book: *Acetylsalicylic Acid*, H. J. M. Barnett *et al.*, Eds. (Raven, New York, 1982) 278 pp.

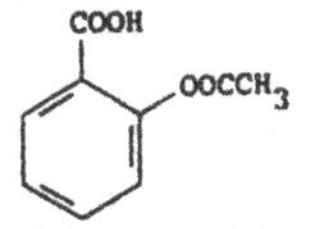

Monoclinic tablets or needle-like crystals. d 1.40. mp 135° (rapid heating); the melt solidifies at 118°. uv max (0.1*N* H_2SO_4): 229 nm ($E^{1\%}_{1cm}$ 484); ($CHCl_3$): 277 nm ($E^{1\%}_{1cm}$ 68). Is odorless, but in moist air it is gradually hydrolyzed into salicylic and acetic acids and acquires the odor of acetic acid. Stable in dry air. pK (25°) 3.49. One gram dissolves in 300 ml water at 25°, in 100 ml water at 37°, in 5 ml alcohol, 17 ml chloroform, 10-15 ml ether. Less soluble in anhydr ether. Decomp by boiling water or when dissolved in solns of alkali hydroxides and carbonates. LD_{50} orally in mice, rats (g/kg): 1.1, 1.5 (Hart).

Guaiacol ester, $C_{16}H_{14}O_5$, *guacetisal, Broncaspin, Guaiaspir.*

Methyl ester, *see* Methyl Acetylsalicylate.

Phenyl ester, *see* Phenyl Acetylsalicylate.

Inorganic salts of acetylsalicylic acid are soluble in water (esp the Ca salt, *q.v.*), but are decomposed quickly.

Pharmaceutical Incompat. (from *Remington's Pharmaceutical Sciences):* Aspirin forms a damp to pasty mass when triturated with acetanilide, phenacetin, antipyrine, aminopyrine, methenamine, phenol or phenyl salicylate. Powders containing aspirin with an alkali salt such as sodium bicarbonate become gummy on contact with atmospheric moisture. Hydrolysis occurs in admixture with salts contg water of crystallization. Solns of the alkaline acetates and citrates, as well as alkalies themselves, dissolve aspirin but the resulting solns hydrolyze rapidly to form salts of acetic and salicylic acids. Sugar and glycerol have been shown to hinder this decompn. Aspirin very slowly liberates hydriodic acid from potassium or sodium iodide. Subsequent oxidation by air produces free iodine.

THERAP CAT: Analgesic; antipyretic; anti-inflammatory.

THERAP CAT (VET): Analgesic; antipyretic; antirheumatic; anticoagulant.

Asphalt, 867
Asphaltum *see* 867
Asphocalcium *see* 855
Aspiculamycin *see* 4436
Aspidin, 868
Aspidinol, 869
Aspidium, 870
Aspidol [Piam] *see* 5510
Aspidosperma, 871
Aspidospermine, 872
Aspirin, 873
Aspirin Aluminum *see* 328
Aspirin Lysine Salt *see* 5510
Aspisol [Bayer] *see* 5510
Aspogen [Eaton] *see* 3168
Aspoxicillin, 874
Aspro [Nicholas] *see* 873
Assaren [Permamed] *see* 3071
Assert [Am. Cyanamid] *see* 4825
Assiprenol *see* 5105
Assugrin *see* 2707
Assur *see* 6181
Assure [Nissan] *see* 8113
Asta C 4898 *see* 3185
Astacene *see* 875
Astacin, 875
Astaril *see* 3807
Astatine, 876
Astaxanthin, 877
Asta Z 4942 *see* 4822
Astemisan [Zdravlje] *see* 878
Astemizole, 878
Astenile [Recordati] *see* 7710
Asteric [Cooper] *see* 873
Asterin *see* 2694

Figure 14.3. Example page from the Merck Index.

Label Reading: Exercise 14a

Look up four chemicals listed on the label of *one* or *two* consumer products that you use. For example, these chemicals might be found in an insecticide, a drug, a food or in some general item like a cosmetic, paint remover, etc. You should have a couple of extra label names with you in case any of your "top four" cannot be found even in the cross index. In addition to *The Merck Index,* you may need to consult one other library reference in answering the mini-essay question. Examples of other references might include:

> *Basic Guide to Pesticides: their characteristics and hazards* by Rachael Carson & Shirley Briggs, © 1992 Hemisphere Pub.
>
> *Complete Drug Reference: United States Pharmacopeia* (1998) Consumer Reports Books. (This is an annual publication.)
>
> *Clinical Toxicology of Commercial Products* by Gosselin, Smith and Hodge, 5th Edition © 1984 by Williams and Wilkins Pub. Classic reference that, unfortunately, has not been updated.
>
> *A Consumer's Dictionary of Food Additives* by Ruth Winter, 5th edition © 1999 by Three Rivers Press.
>
> *Food Additives, Nutrients & Supplements A - Z* by Eileen Renders, © 1999 by Clean Light Publishers.
>
> *General Science Index,* James Kochones, Editor, © H. W. Wilson Co. (This is an ongoing, multivolume guide to periodicals—the science version of *Reader's Guide to Periodical Literature*.)

Label Reading: Exercise 14b

You will look up some properties of eight chemicals found in the four categories of products described below. When using *The Merck Index,* do not expect to find information for the last two blanks (side effects and toxicity) for every compound. Look at the examples given on the report sheets. When making a list of ingredients to look up in *The Merck Index,* it is smart to include some extras beyond the minimum of eight; twelve might be a safer number to start with. There will usually be some chemicals that you will not be able to locate even in the cross index.   In such cases, don't become frustrated—simply choose an alternate ingredient from your list and proceed.

The categories of ingredients that you will need to research are:

A. "Cides": (Insecticides, pesticides, fungicides, herbicides, algaecides, rodenticides). All these are killer chemicals. Locate one or more "cide" products containing a total of two different active ingredients.

B. Drugs: Locate one or more pharmaceutical/medicinal products containing a total of two different active ingredients.

C. Consumer Products: (Things that you use in your home, kitchen, garage, or on yourself—such as cosmetics, cleaners, paint removers, polishes etc.). The labels may prove somewhat skimpy on chemical details in this category, but you should be able to come up with one or more products containing a total of two different ingredients.

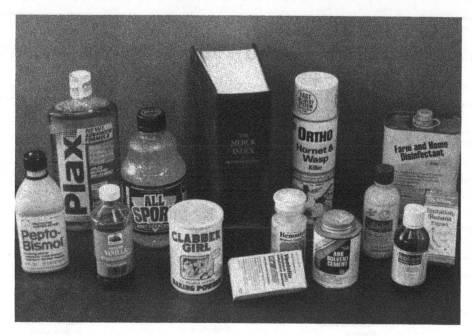

Figure 14.4. The Merck Index and some consumer products.

D. Food Additives: Locate food products containing a total of two different additives found among the eight categories listed below:

Anticaking agents   *Sequestrants*
Chemical preservatives   *Stabilizers*
Emulsifying Agents   *Synthetic flavoring substances*
Nutrients & dietary supplements *Multiple purpose & misc. additives*

A partial listing of chemicals used in each of these food additive categories appears in the *Generally Recognized as Safe* ("GRAS") list on the next page.

To facilitate your search and get you started, one entry has already kindly been provided in each category on the report sheets as a guide. This leaves only two more ingredients that you must look up in each category. The total of all compounds shown on the report sheets will be twelve—four already provided and eight researched by you.

Direct Food Substances Generally Recognized as Safe
(a partial listing of additives in Food & Drug Administration's "GRAS" list)

Anticaking Agents

Aluminum calcium silicate
Calcium silicate
Magnesium silicate
Sodium aluminosilicate
Sodium calcium aluminosilicate
Tricalcium silicate

Chemical Preservatives

Ascorbic acid
Ascorbyl palmitate
Benzoic acid
Butylated hydroxyanisole
Butylated hydroxytoluene
Calcium ascorbate
Calcium propionate
Calcium sorbate
Dilauryl thiodipropionate
Erythorbic acid
Methylparaben
Potassium bisulfite
Potassium metabisulfite
Potassium sorbate
Propionic acid
Propyl gallate
Propylparaben
Sodium ascorbate
Sodium benzoate
Sodium bisulfite
Sodium metabisulfite
Sodium propionate
Sodium sorbate
Sodium sulfite
Sorbic acid
Stannous chloride
Sulfur dioxide
Thiodipropionic acid
Tocopherols

Emulsifying Agents

Diacetyl tartaric acid esters
    of mono- and diglycerides
Mono- and diglycerides of fats and oils
Monosodium phosphate derivatives
    of the above
Propylene glycol
Ox bile extract

Nutrients and Dietary Supplements

Ascorbic acid
Biotin
Calcium carbonate
Calcium citrate
Calcium glycerophosphate
Calcium oxide
Calcium pantothenate
Calcium phosphate
Calcium pyrophosphate
Calcium sul fate
Carotene
Choline bitartrate
Choline chloride
Copper gluconate
Cuprous iodide
Cysteine
Ferric phosphate
Ferric pyrophosphate
Ferric sodium pyrophosphate
Ferrous gluconate
Ferrous lactate
Ferrous sulfate
Inositol

Iron, reduced
Linoleic acid
Magnesium oxide
Magnesium phosphate
Magnesium sulfate
Manganese chloride
Manganese citrate
Manganese gluconate
Manganese glycerophosphate
Manganese sulfate
Manganous oxide
Niacin
Niacinamide
D-pantothenyl alcohol
Potassium chloride
Potassium glycerophosphate
Potassium iodide
Pyridoxine hydrochloride
Riboflavin
Riboflavin-5-phosphate
Sodium pantothenate
Sodium phosphate
Sorbitol
Thiamine hydrochloride
Thiamine mononitrate
Tocopherols
Tocopherol acetate
Vitamin A
Vitamin A acetate
Vitamin A palmitate
Vitamin B_{12}
Vitamin D_2
Vitamin D_3
Zinc chloride
Zinc gluconate
Zinc oxide
Zinc stearate
Zinc sulfate

Sequestrants

Calcium acetate
Calcium chloride
Calcium citrate
Calcium diacetate
Calcium gluconate
Calcium hexametaphosphate
Calcium phosphate, monobasic
Citric acid
Dipotassium phosphate
Disodium phosphate
Isopropyl citrate
Monoisopropyl citrate
Potassium citrate
Sodium acid phosphate
Sodium citrate
Sodiuum diacetate
Sodium gluconate
Sodium hexametaphosphate
Sodium metaphosphate
Sodium phosphate
Sodium potassium tartrate
Sodium pyrophosphate
Sodium pyrophosphate, tetra
Sodium tartrate
Sodium thiosulfate
Sodium tripolyphosphate
Stearyl citrate
Tartaric acid

Stabilizers

Acacia (gum arabic)
Agar-agar
Ammonium alginate
Calcium alginate
Carob bean gum
Chondrus extract

Ghatti gum
Guar gum
Potassium alginate
Sodium alginate
Sterculia (karaya) gum
Tragacanth

Synthetic Flavoring Substances

Acetaldehyde
Acetoin
Anethole
Benzaldehyde
N-butyric acid
d- or l-carvone
Cinnamaldehyde
Citral
Decanal
Diacetyl
Ethyl acetate
Ethyl butyrate
Ethyl vanillin
Eugenol
Geraniol
Geranyl acetate
Glycerol tributyrate
Limonene
Linalool
Linalyl acetate
Methyl anthranilate
3-Methyl-3-phenylglycidic
    acid ethyl ester
Piperonal
Vanillin

Multiple Purpose and Miscellaneous Additives

Acetic acid
Aconitic acid
Adipic acid
Alginic acid
Brown algae
Red algae
Aluminum ammonium sulfate
Aluminum potassiun sulfate
Aluminum sodium sulfate
Aluminum sulfate
Ammonium bicarbonate
Ammonium carbonate
Ammonium chloride
Ammonium hydroxide
Ammonium phosphate, monobasic
Ammonium phosphate, dibasic
Ammonium sulfate
Bakers yeast extract
Beeswax (yellow and white)
Bentonite
n-Butane and isobutane
Caffeine
Calcium hydroxide
Calcium iodate
Calcium lactate
Calcium stearate
Candelilla wax
Caprylic acid
Caramel
Carbon dioxide
Carnauba wax
Mixed carbohydrase and
        protease enzyme product
Carngluten
Clove and its derivatives
Cocoa butter substitute
    primarily from palm oil

Copper sulfate
Corn silk and corn silk extract
L-Cysteine monohydrochloride
Dextrin
Dextrans
Diacetyl
Dill and its derivatives
Ethyl alcohol
Ethyl formate
High fructose corn syrup
Garlic and its derivatives
Insoluble glucose isomerase
    enzyme preparations
Glutamic acid
Glutamic acid hydrochloride
Glycerin
Glyceryl monostearate
Helium
Hydrochloric acid
Hydrogen peroxide
Lactase enzyme preparation
    from kluyveromyces lactis
Lactic acid
Lecithin
Ground limestone
Maltodextrin
Malt syrup (malt extract)
Methylparaben
Magnesium carbonate
Magnesium hydroxide
Magnesium stearate
Malic acid
Methylcellulose
Monoammonium glutamate
Monopotassium glutamate
Nickel
Nitrogen
Nitrous oxide
Ozone
Panain
Pectins
Peptones
Phosphoric acid
Potassium acid tartrate
Potassium bicarbonate
Potassium carbonate
Potassium hydroxide
Potassium sulfate
Potassium iodate
Propane
Pyridoxine hydrochloride
Rennet (animal derived)
Rapeseed oil
Rue
Oil of rue
Silica aerogel
Sodium acetate
Sodium aluminum phosphate
Sodium bicarbonate
Sodium carbonate
Sodium carboxymethylcellulose
Sodium caseinate
Sodium hypophosphite
Sodium hydroxide
Sodium pectinate
Sodium phosphate
Sodium sesquicarbonate
Stearic acid
Succinic acid
Sulfuric acid
Tartaric acid
Triacetin
Triethyl citrate
Urea
Wheat gluten
Whey
Reduced lactose whey
Reduced minerals whey
Whey protein concentrate
Zein

Label Reading

Date _____   Section number _____   Name _____

Name of consumer item(s)   _____.

    Chemical ingredient #1: _____.

        alternate name   _____

        chemical formula   _____.

        uses   _____.

        LD_{50} or MLD (if listed)   _____.

    Chemical ingredient #2: _____.

        alternate name   _____.

        chemical formula   _____

        uses   _____

        LD_{50} or MLD (if listed)   _____.

    Chemical ingredient #3: _____.

        alternate name   _____.

        chemical formula   _____.

        uses   _____

        LD_{50} or MLD (if listed)   _____.

    Chemical ingredient #4: _____.

        alternate name   _____.

        chemical formula   _____.

        uses   _____

        LD_{50} or MLD (if listed)   _____.

Label Reading

Date _____   **Section number** _____   **Name** _____

1. We read and hear of advertisers claiming that *their* product contains "no chemicals."

   (a) If this were *really* true, what would be inside the product package?

   (b) What is the manufacturer *really* trying to say?

2. Write a 200 word description on some aspect of *one* of your four ingredients listed on the report sheet. This mini-essay can discuss discovery, manufacture, history, uses, dangers, addiction, your experience, etc.

Reference Used _____

Label Reading

Date _____   Section number _____   Name _____

(A) "Cides"	Brand and Product Name	Ingredient Name	Alternate Chemical Name	Formula	Uses	Side Effects: Contra-indica-tions	Toxicity: LD$_{50}$ or MLD
(1) Rodenticide	Havoc	Brodifacoum	Talon	$C_{31}H_{23}BrO_3$	kills mice and rats	———	0.27 mg/kg in rats
(2)							
(3)							

Label Reading

Date _____   Section number _____   Name _____

(B) Drugs	Brand and Product Name	Ingredient Name	Alternate Chemical Name	Formula	Uses	Side Effects: Contra-indica-tions	Toxicity: LD_{50} or MLD
(1) Sheep and horse wormer	Eqvalan paste	Ivermectin	22,23-di-hydroaver-mectin	$C_{48}H_{74}O_{14}$	Anti-parasitic	Do not use in dogs	----------
(2)							
(3)							

Label Reading

Date _____   Section number _____   Name _____

(C) Consumer Products	Brand and Product Name	Ingredient Name	Alternate Chemical Name	Formula	Uses	Side Effects: Contra-indications	Toxicity: LD 50 or MLD
(1) Paint remover	Zip-Strip	Methylene chloride	Dichloro-methane	CH_2Cl_2	Solvent; degreasing fluid	Narcotic	————
(2)							
(3)							

Label Reading

Date _____   Section number _____   Name _____

(D) Food Additives	Brand and Product Name	Ingredient Name	Alternate Chemical Name	Formula	Uses	Side Effects; Contra-indica-tions	Toxicity: LD$_{50}$ or MLD
(1) Anti-caking agent	Leslie iodized salt	Aluminum calcium silicate	Calcium alumino-silicate	CaAl$_2$Si$_2$O$_8$	Food; also used in cements	-------	-------
(2)							
(3)							

Experiment 15

Chromatography of Artificial Colors

Samples From Home

No samples from home are need for this experiment.

Objectives

This experiment introduces you to the world of artificial colors by allowing you to separate a solution that appears to be a single component into the separate dyes that act together to give that mixture its color. The technique of separation used in this experiment is called liquid chromatography.

Background

Hundreds of thousands of pounds of artificial colors are added to food, drugs and cosmetics each year in the United States. This artificial coloring is used solely to enhance the visual appeal of the product to the consumer. Artificial colors are regulated in the United States by the Food and Drug Administration (FDA) under the Food, Drug and Cosmetics Act. Regulation in this area was first instituted in 1906 when seven dyes were allowed. One of these, "Butter Yellow, was banned in 1932 when research showed that it caused liver tumors in rats. By 1950, the list of approved dyes had expanded to nineteen. In that year, three of those approved dyes were removed from the list when children eating popcorn colored with them became sick. More recently, FD&C (Food, Drug, and Cosmetic) Red 2 was banned by the Food and Drug Administration in 1976, again based upon cancer concerns. Presently there are seven dyes approved for use in food in the United States and a few more approved for dying things like orange skins (Citrus Red No. 2) that are not supposed to be consumed. Citrus Red No. 2 is listed by the State of California as a known carcinogen.

Many consumer products have more than one added artificial dye. Many popular "fruit" drinks, for instance, a staple in American childhood, often have two and sometimes three or more dyes added to yield those pleasing bright colors that make them so attractive to kids. In this experiment you will separate a few of the dyes that have been added to some consumer products using a small plastic cartridge called a Sep Pak® and a separation technique called liquid chromatography.

The purpose of all chromatography is to separate mixtures into individual components. Liquid chromatography gets its name from the procedure of passing a liquid mixture through a porous solid (called the stationary phase) that has been packed into a (usually) metal column. This mixture is dissolved in a solvent, called the mobile phase, that is pumped continually through the chromatographic column. As the liquid components in the mixture pass through the column, they are attracted by differing amounts to the stationary phase packed in the column. Actually, they go through repeated cycles of "dissolving" in the stationary phase then redissolving into the moving mobile phase and moving down the column again to some degree before interacting with the solid phase again. This cycle occurs over and over again. Since the way or the amount that each component in the mixture does this is slightly different (that is, more or less interaction) the components become separated from each other by the time they reach the column's end. The end of the column is where the separated components are detected or in our case collected for further analysis.

After we give you the tools and techniques for this simple chromatographic experiment, you will be asked to design mixtures (of the two solvents used) that change the separation ability (chromatography) of the Sep Pak cartridge. Pay attention to what you do in each step, and the further modifications necessary to clearly separate the components in your mixture won't be very hard. The key to this lab is experimentation, evaluation of the results, and then additional experimentation. In a word: chemistry.

Procedure

Caution: *Methanol is poisonous and flammable. Use care when handling this solvent.*

1. Line up 10 test tubes (150 mm; 6 inch) against a white background in your test tube rack. Fill a 100 mL beaker with water and another with methanol. Mark these beakers with labels or a grease pencil. Methanol is less polar than water. You can mark the water beaker with a *P* if you wish.

2. Draw 10 mL of methanol into a 25 mL syringe. This is 10 mL of 100% methanol. Attach the syringe to the *short end* of the Sep Pak cartridge. Almost any 25 mL or larger disposable syringe with removable needles will fit. Flush the Sep Pak cartridge with the methanol by pressing the plunger slowly all the way to the bottom. Direct the *methanol* that is exiting (eluting) from the cartridge into an appropriate waste beaker. 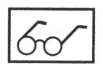 Press the syringe plunger down in such a way that approximately 3 drops per second come out of the end of the cartridge. You cannot depress the plunger too slowly; however, if you

press too hard liquid will leak out of the connection between the syringe and the cartridge or it will squirt out from behind the plunger back up the barrel of the syringe. Careful!

3. Repeat the flushing process by drawing 5 mL of *water* into the syringe with the cartridge disconnected and then passing all of the water through the reconnected cartridge. Disconnect the cartridge from the syringe and get 50 mL of an unknown dye mixture from the lab instructor. RECORD YOUR UNKNOWN NUMBER AND ITS INITIAL COLOR ON THE REPORT SHEET.

4. Draw 5 mL of your unknown dye mixture into the syringe. Reconnect the syringe to the short end of the cartridge, and slowly pass the mixture through the cartridge into the waste beaker. The liquid eluting (exiting) from the cartridge will be relatively clear, while the dyes in the unknown mixture will stay on the cartridge. This is called charging the column.

5. Experimentally you now need to design a scheme using different mixtures of methanol and H_2O that will allow you to correctly determine the total number of different color dyes in your unknown mixture. Total volumes of 10 mL are probably best for any mixture of solvents that you choose to use. Keep a listing of the mixtures that you use in case you need to repeat or refine your procedure. For example, if you want to start with a 100% water "mixture," then draw 10 mL of water in the syringe, connect the charged cartridge and slowly pass that solvent through the cartridge, collecting the liquid in a test tube that is eluting from the cartridge. Next try a 90% water/ 10% methanol mixture. This can be made by drawing 9 mL of water into the syringe and then without expelling the water drawing 1 mL of methanol into the syringe. This solution can then be successfully mixed by drawing a 3 mL air bubble into the syringe and inverting the syringe repeatedly. Make sure that you get rid of the air bubble before you start to pass the solvent onto the cartridge.

As you elute these water/methanol mixtures (that is, pass them one by one through the syringe and cartridge), individual colored dyes will start to come off of your chromatographic column (the cartridge). Collect these in test tubes as they exit from the cartridge. You may be able to see the colored bands as they separate and move down the cartridge since the wall of the Sep Pak is a relatively thin, white plastic.

6. If you see a change in the color of the solvent eluting from the syringe, collect it in another test tube. If it becomes obvious that you have mixed two colors and you want to begin again, cleanup by passing 20 mL of methanol and then 5 mL of water through the cartridge (into the waste beaker) and starting again at Step 4.

7. Repeat this process until you have as many test tubes as possible with different colors in them. Try your best to make each test tube as monochromatic (containing only one color) as possible. For instance, a green solution in a test tube is probably a poorly separated mixture of pure blue and pure yellow dyes. Similarly, a purple solution is probably unseparated red and blue dyes. Think back to the color wheel in grammar school to decide which primary colors combine to make which secondary colors.

8. Don't be timid in the design of your elution scheme. If one thing doesn't work, completely cleanup the column as described in Step 6 and start again. Some of the unknown dye mixtures may contain dyes that are more difficult to separate than others. Don't be intimidated by the

success or failure of your peers; maybe their unknown mixture has a different dye combination than yours. Experiment! After all, this is the foundation upon which the field of chemistry is built. Record the final optimized solvent elution scheme on the report sheet.

Chromatography of Artificial Colors

Date _____ **Section number** _____   **Name** _____

Unknown number and initial color_____

List the individual colors (dyes) determined from your final (optimized) elution scheme:

_____

_____

_____

_____

_____

List the solvent mixtures that you used in your longest, most successful elution (optimized) scheme:

<u>Example:</u>   <u>60% water/40% methanol</u>

_____

_____

_____

_____

_____

_____

_____

Chromatography of Artificial Colors

Date _____   **Section number** _____   **Name** _____

1. Which is a more polar solvent, methanol or water?

2. What is the relative polarity of the stationary phase in your Sep Pak cartridge, polar or nonpolar?

3. Why are artificial colors used in consumer products?

4. What would happen to manufacturers' profits if artificial colors were banned in the United States?

5. What are the relative polarities of the dyes encountered in this experiment? List the dyes that you separated (colors) from most polar to least polar. This should be the relative order in which they eluted from your cartridge as you added more and more methanol to your elution mixtures. See the experimental procedure to remind yourself which is more polar, water or methanol.

Think, Speculate, Reflect, and Ponder

6. Why does the procedure for this experiment suggest beginning the elution scheme with a 100% water solution and adding increasingly more methanol instead of starting with a 100% methanol solution and adding increasingly more water?

7. If the stationary phase in a liquid chromatographic column is made of very polar silica, how can a nonpolar stationary phase be made without using a new stationary phase? (Hint: Try looking in chromatography or quantitative analysis texts.)

Experiment 16

Warning: This Experiment May Contain Lead

Samples From Home

Bring ONE of the following to check for lead in paint chips or pottery glaze:

1. A small amount of a substance that might contain lead such as old paint chips. You will need approximately 0.2 g; if a liquid sample is chosen, it must be a water-based solution.

2. If the lead content of a pottery glaze is to be examined, you must bring *two* liquid samples—at least 50 mL of white distilled vinegar directly from the bottle AND a 100 mL volume of the same vinegar which has been allowed to soak in an earthenware vessel for 24 hours at room temperature.

Objectives

A sample from home will be analyzed both *qualitatively* (what is present) and *quantitatively* (how much of what is present) for lead content. Paint chip analysis will introduce the techniques of washing, centrifuging, handling small samples and the use of an accurate type of analytical balance. An alternative analysis procedure for pottery glaze leachate solutions will use a spectrophotometer to demonstrate how light can be used as a quantitative tool. These results should permit a valid judgment regarding the safety hazard that a particular sample may pose by comparison to maximum permitted lead concentrations.

Background

Lead (elemental symbol Pb from the Latin *Plumbum*) is a metal that has been known since very ancient times because it is easy to liberate the free metal from its ore. Lead melts at a relatively low temperature of 327 °C and is a heavy (dense), relatively unreactive element. It has been used to protect (paints, cable coverings), propel people (car batteries, leaded gasoline) and, of course, shoot people (bullets). Its role and effect upon our lives is a mixed blessing, and in this respect it is therefore no different from most chemicals. Used appropriately, chemicals can enhance our lives; used unwisely, man and this planet may suffer the consequences.

It has been only during relatively recent times, however, that human beings have grown in their appreciation and respect for some of the more subtle and insidious problems associated with the use of chemicals previously thought of as safe. Atoms, of course, are not anthropomorphic (exhibiting human characteristics) and, therefore, know neither friend nor foe. Instead, atoms respond only to the laws of nature which human beings, try as they might, cannot alter. Lead is a member of the so-called *heavy metal poison club*, the full extent and nature of whose toxicities are only now slowly coming to light. You have probably heard of other prominent members, notably mercury, cadmium and arsenic.

Mercury, associated with industrial water pollution and the indiscriminate use of certain fungicides, can have a terrible effect upon the brain and nervous system. Cadmium is found in industrial water pollution, earthenware glazes and iron water pipes; its alleged implication in heart disease is still being investigated. Arsenic, a suspected strong cancer inducing agent and very toxic element, has received perhaps the most notoriety from the classic movie *Arsenic and Old Lace* and as a by-product from the refining of arsenic containing copper ores.

As with other heavy metals, lead is especially dangerous due to its ability, even in extremely low doses, to accumulate in soft bone tissue in the body much faster than the body can excrete it. Because of this, it is termed a *cumulative* poison and, once present in the body, it is a medically difficult and dangerous job to get rid of it. Some of the more celebrated sources of lead pollution in our environment are large scale air

Experiment

and dust contamination from copper smelters (where lead is often an impurity or minor constituent in the ore) and certainly from 70 years of burning leaded gasoline containing TEL. Seven million tons of this **t**etra **e**thyl **l**ead compound were added to gasoline from 1923 to phase out 1986. The installation of particle precipitators in smelters' smoke stacks and the conversion to unleaded gasoline for cars have done much to reduce waste lead pollution. The success of these changes can be seen by the precipitous drop in lead dust emissions over just a 17 year period (Table 16.1).

Lead Emissions (standard tons) in the U.S.		
Year	Road Vehicles	Total Emissions
1970	172,000 Tons	224,000 Tons
1980	62,000 Tons	78,000 Tons
1987	3,100 Tons	8,900 Tons

Table 16.1.**Lead Emissions**.

Persuasive evidence has appeared supporting the fascinating contribution of lead poisoning to the fall of the Roman Empire. Lead-based plumbing, lead lined wine casks and expensive lead salts for "sweetening" wine—available only to upper class, ruling Romans—may have contributed significant lead doses to their bodies. And similar to those "rich" Romans, there has long been concern that lead solder in water pipes could leach into our water supply. For this reason it is still probably advisable (1) not to use water from the hot water line for drinking, and (2) to let the cold water run for a couple of minutes the first time it is turned on each morning.

In the past, school drinking fountains and faucets were even shut off after it was discovered that lead concentrations in the water from lead solder or galvanizing often greatly exceeded acceptably safe levels. Since 1991, the U.S. Environmental Protection Agency has enforced a maximum permitted lead level in household water supplies of 15 ppb (parts per billion). Lead test kits are available in stores for around $20 which claim to detect lead concentrations in water exceeding this EPA limit.

And the Sunday comics might not seem so funny when one realizes that until 1986, their colored inks sometimes contained lead which was released into the environment by burning (even putting the ashes on your garden was inadvisable). There is a good chance that you have read about leaded paints—especially old and peeling—being eaten by children, or paint dust being inhaled during housing renovations. Fortunately, a better understanding of the long term effects of lead on health has resulted in federal regulations requiring all paints manufactured for the general public since 1978 to be lead-free. But the legacy of misuse of this cumulative poison lead over past generations remains. (Check out the fascinating effect of lead in "Gout and Genius" on page 1015 in the December, 1994, issue of *Journal of Chemical Education*.)

In October, 1991, the U.S. Centers for Disease Control lowered the threshold for lead poisoning to 10 micrograms lead per 1 deciliter of blood (0.00001%). Current estimates suggest that one million children living in the U.S. still exceed this lead level, putting them at risk of irreversible damage to their health. A child ingesting less than one sugar size granule of lead paint dust each day would reach blood lead levels exceeding this amount. Even though lead in paint is now banned for consumer use, 74% of all private housing built before 1980 contains some lead paint.

Finally, there was the discovery (after some unfortunate poisonings) that the lead in pottery glazes, if improperly fired (heated), can be leached out by food stored in them -- especially acidic fruit juices. This dose of lead is thereby passed on into the unsuspecting consumer. As monitored and enforced by the U.S. Food and Drug Administration, the current 2002 maximum leachable lead levels permitted in earthenware are 0.00005 to 0.0003% (0.5-3 ppm) depending upon the type and dimensions of the vessel. Acceptable cadmium levels (not measured in this experiment) are 0.5 ppm.

Depending upon your interests and particular sample chosen, two *qualitative/quantitative* procedures are described. The **first method** determines lead *gravimetrically* (the weighing of precipitates) and handles lead concentrations of 0.1% or greater in solid samples like paint chips. Concentrations down to 0.01% Pb in liquid samples can be detected *qualitatively* (detected but the concentration not measured) by this method. (For those getting a positive lead test, an additional, but optional, *quantitative* procedure is available from your lab instructor.) A **second method** examines the amount of lead *spectrometrically* (in this case how precipitates scatter and block the passage of light through a liquid) and it is mainly applicable for pottery used to store food and liquids. This spectrometric method can detect the very small amounts of lead that may be present in these pottery glaze leach solutions (0.0005 to 0.0010% Pb).

The chemistry of the **gravimetric method** begins by using nitric acid to decompose your solid sample to dissolve out the lead in the form of lead ions. This lead is then precipitated by adding potassium iodide which makes the positive lead ions and negative iodide ions combine to form an insoluble lead iodide precipitate as indicated in the chemical reaction:

$$Pb(NO_3)_2 \quad + \quad 2KI \quad \longrightarrow \quad 2KNO_3 \quad + \quad PbI_2$$

| lead (II) nitrate (soluble) | potassium iodide (soluble) | potassium nitrate (soluble) | lead (II) iodide (insoluble) |

Or, using what is called a net ionic equation, this can be written simply as:

$$Pb^{+2} \quad + \quad 2I^{-1} \quad \longrightarrow \quad PbI_2$$

lead ions   iodide ions   lead (II) iodide precipitate

Of the possible metals likely to be present, only lead will form an insoluble iodide precipitate. (Sodium sulfite is also added to this solution to reduce unwanted oxidation of the iodide ion by the nitric acid present.)

If you choose to continue with a *quantitative* determination at this point, you will wash this precipitate with water to remove any of the soluble compounds sticking to it, followed by a final wash with alcohol to wash away the water itself and permit a more rapid drying of the precipitate. Such preparation of a sample for accurate weighing is the nature of good and necessary chemical technique. Since this method will give a yellow PbI_2 precipitate down to 0.01% Pb in your 2 mL of solution, this means that any solid sample containing 0.2 milligram (0.0002 g) of lead or more should give a positive test. For a 0.20 g sample, this would mean that a lead concentration as low as 0.1% (1000 ppm) Pb will be detectable.

The spectrometric method (used with leaded pottery glazes) precipitates the lead as lead chromate:

$$Pb(NO_3)_2 \quad + \quad K_2CrO_4 \quad \longrightarrow \quad KNO_3 \quad + \quad PbCrO_4$$

lead (II) nitrate (soluble)   potassium chromate (soluble)   potassium nitrate (soluble)   lead (II) chromate (insoluble yellow precipitate)

This procedure uses chromate instead of iodide ions to precipitate the lead. While not so selective (some other metals besides lead, such as barium and copper, will also precipitate as chromates), lead chromate is one of the most insoluble of all ionic compounds, only 0.0002 g dissolving in one liter of pure water. By contrast, the "insoluble" lead iodide dissolves to the extent of about 0.7 g per liter—still quite insoluble, but nevertheless over 1000 times more soluble than the lead chromate! Chemists understand that soluble and insoluble are only relative terms, and that virtually nothing is *completely* insoluble in water.

The procedure calls for adding sucrose (sugar) to the solution prior to precipitation. This chemical acts as a stabilizing agent by controlling the growth of the precipitating particles. This makes for a smaller and more uniform particle size while retarding the particles from settling out of the solution. In the concentrations used it can also serve to reduce the solubility of the lead chromate still further. Such physical aspects of a precipitate are important when trying to measure the amount of a solid present based on the turbidity (cloudiness) of the solution. The greater the turbidity, the more precipitate present and thus the higher the amount of lead in your sample.

A *quantitative* measure of this turbidity is performed using a light beam; that is the function of the Spectronic 20 spectrophotometer. This instrument shines specific wavelengths of light through your sample and electronically detects how much of this light is able to pass through your solution by recording the light not scattered or otherwise absorbed along the way by your lead chromate precipitate. This instrument is thus comparing the amount of light passing *out of* (exiting) the solution to the light passing *into* (entering) the solution. This ratio of light exiting divided by the light entering gives us the *light transmittance* numbers that you read off the instrument and the lead calibration graph. (The numbers have been multiplied by 100 to convert to percent.) A 50% transmittance, for example, means that only half the light is able to pass through the solution.

Your light transmittance data from the lead sample must next be compared to that for a zero percent lead sample (the "blank") so that the final lead concentration can simply be read off the calibration graph shown later in this experiment. Locate the average *transmittance* value for your samples on this graph, read horizontally right over to the heavy, black calibration line; then read vertically down to get the corresponding *ppm lead* value. (Very accurate measurements would actually require each group to make their own calibration graph using their own vinegar solution.)

Procedure

A. Gravimetric Method *(for solid samples like paint chips)*

1. Weigh out, to the nearest 0.01 g, about 0.2 g of a solid (or 2.0 g of an aqueous liquid) sample to be tested and place it into a 3 inch (10 x 75 mm) test tube. If a liquid sample is used, add 3 drops of dilute nitric acid and proceed directly to procedure Step 2. Otherwise, add 1 mL (about 20 drops) of dilute nitric acid [6*M* HNO$_3$] to the solid sample and heat for 15 minutes by placing the test tube into a 100 or 150 mL beaker half full of boiling water. If your sample contains any chunks which do not break up in the acid, use a short solid glass rod to mash them up. Do not lay the rod down on the desk top, but leave it in the test tube. (This prevents "losing" some of your sample on the bench top.)

Good technique is what separates the careful, meticulous and successful worker from one who is not successful. The liquid in your test tube contains possible soluble lead ions at this point. You must try not to drop, spill, or otherwise "lose" any of this liquid, some of which is now on your stirring rod,

Figure 16.1. Heating paint chip in a 3 inch test tube using a boiling water bath.

Figure 16.2. A small centrifuge.

and some of which will shortly be inside a medicine dropper. Chemists avoid losing any sample by both not spilling and by "washing it all out." That's where we're at now as the experiment continues.

Return to your sample heating in the water bath and pull the stirring rod about half way up the inside of the tube and wash it off with 10 drops of distilled water so that the washings fall and mix in with your sample. Set the rod aside. Centrifuge the test tube and contents for a minute. Your lab instructor will demonstrate the use of the centrifuge: remember to use a counter weight tube for proper balance.

Using a long-nosed medicine dropper, carefully—technique is important here—transfer as much liquid as you can into a 3 inch test tube (liquid only). If the solids get stirred up, you can always centrifuge again. When you have transferred as much liquid as feasible without sucking up any solid, add 10 more drops of distilled water to your test tube containing the solid and repeat the liquid transfer technique of centrifuging and sucking off the liquid with the dropper. Combine the wash liquid with the original solution. When done properly, all of the solid, now washed free of lead ions, will be left in your original tube, while a total of about 2 mL of clear (but possibly colored) liquid containing almost all of your goodies is now in your weighed tube.

2. Add 1 drop of sodium sulfite solution [$1M$ Na_2SO_3] to your liquid sample. Mix by pressing a small piece of plastic wrap over the tube mouth with your finger and inverting the tube twice. Concentrations of lead much higher than 10% in your original solid sample (or 1% in the larger liquid sample) may result in the formation of a persistent cloudiness at this point, but this will not hurt your results.

(a) Add 3 drops of potassium iodide solution [$2M$ KI] and mix well by inverting the test tube as before. What do you observe? A yellow precipitate (not just a colored solution) indicates about 0.1% lead or more in your solid sample, or 0.01% or more if you started with two grams of a liquid sample.

(b) ONLY if you obtained negative results in step 2(a), add a drop of lead nitrate solution [$0.1M$ $Pb(NO_3)_2$] to the test tube containing the sample to see what a positive test looks like.

3. If your test from Step 2(a) is positive and you wish to do a quantitative determination, see your lab instructor for that procedure (extra time needed is about 60 minutes).

B. Spectrometric Method *(for vinegar leach samples)*

1. NOTE: All glassware coming into contact with your solutions to be tested must be previously scrubbed with soap and water and then rinsed with distilled water. You should also have with you two white distilled vinegar or 5% acetic acid solutions. One has come "right from the bottle" (label this one *blank*), while the other has been allowed to stand in some earthenware vessel for 24 hours at room temperature (label this one *leach*). If either or both of these solutions is not clear of any turbidity, they must be filtered or centrifuged before proceeding further. And if any color is

present in your leach solutions, they will have to be clarified. See your lab instructor for help in the unlikely event that your solution is either cloudy or colored.

Figure 16.3. Preparing blank and leach samples for lead analysis.

Pour 35 mL of the *blank* solution into one 125 mL Erlenmeyer flask and two 35 mL portions of the *leach* solution into each of two 125 mL Erlenmeyer flasks. You should now have a total of three flasks each containing 35 mL of liquid—one blank and two duplicate leach samples. Add to *each one* of these three flasks 3.5 g sucrose and swirl the contents of each flask until the solid has dissolved.

2. While continuously swirling your blank solution, rapidly add 1.00 mL of dilute potassium chromate [$0.5M$ K$_2$CrO$_4$] from a pipette with a pipetter bulb attached. (A type like the

"Propipetter" works well.) Your lab instructor can indicate the proper and safe technique for using a pipette.) Continue swirling without stopping for 1 minute. (If pipettes are not available, add rapidly, with swirling, 5 drops of concentrated potassium chromate [$2M$ K$_2$CrO$_4$] using a stubby medicine dropper.) Repeat this identical procedure with the two leach solutions and allow all three flasks to stand for at least 10 minutes. Note your observations on the report sheet. Any haziness in your leach solutions, however faint, indicates the presence of at least 5 ppm (0.0005%) Pb in your sample.

3. Pour each of your solutions into three separate matched spectrophotometer *cuvettes* (precision made test tubes). Your lab instructor can show you how to "match" cuvette tubes yourself, if necessary. With no sample tube in the Spectronic 20 instrument, set the transmittance at 0% with the *zero control knob.* Then wipe the outside of the cuvette containing the *blank* with a tissue and insert it into the sample holder in the top of the instrument so that the vertical mark on the top directly faces you. Close the instrument's cover cap.

Check that the wavelength dial is set on 540 millimicrons, mµ (540 nanometers, nm)—adjust if necessary now, but do not touch the wavelength knob again throughout the rest of your readings. Using the *light control knob,* adjust until the transmittance reads 100%. (If you are among the first to use the instrument, your lab instructor will probably have you recheck and read just the zero and percent transmittance settings until the instrument completely warms up.)

Figure 16.4. Pipette addition of potassium chromate to sample.

(a) Your instrument is now standardized against your pure vinegar (or the 5% acetic acid reference blank) containing 0% lead. Remove the *blank* cuvette and insert the *leach* cuvette solution into the sample holder, wiping and aligning the tube as described before.

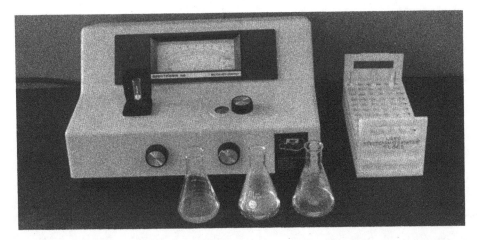

Figure 16.5. Spectrophotometer (turbidimeter) for analyzing samples.

Record the % transmittance reading. Repeat this procedure with your other duplicate leach solution. Faint cloudiness in the cuvette tube solutions can best be seen visually by looking down on the top of the tube and thus through the whole length of solution while holding the tube against a black background.

(b) Estimate the concentration in parts per million (ppm) of leachable lead in your two leach solutions (10 ppm = 0.001% = 10 $^{mg}/L$ of lead) by referring to one of the calibration curves on the next page. Make sure that you use the correct curve for your samples. The closer your lead values are in the duplicate samples, the more precise you have been in handling both samples in an identical manner.

If your results are negative (no detectable lead), you can see what a positive test looks like by adding a drop of lead nitrate solution [0.1 M Pb(NO$_3$)$_2$] to one of your two sample tubes.

If you have too much lead in your samples, then the percent transmittance (%T) will be so low that the lead concentration cannot be read off the graph. In this case, you will need to make a dilution of your leach solution with pure vinegar before testing. If time permits, see your lab instructor regarding how to do this and what dilution factor to use.

Lead Calibration Curves (Graphs)
5% Vinegar Leach Solution

5 % Vinegar Solution
Pipette Addition of K_2CrO_4

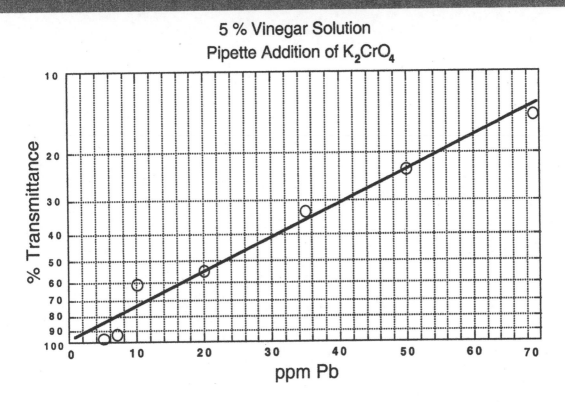

5 % Vinegar Solution
Dropwise Addition of K_2CrO_4

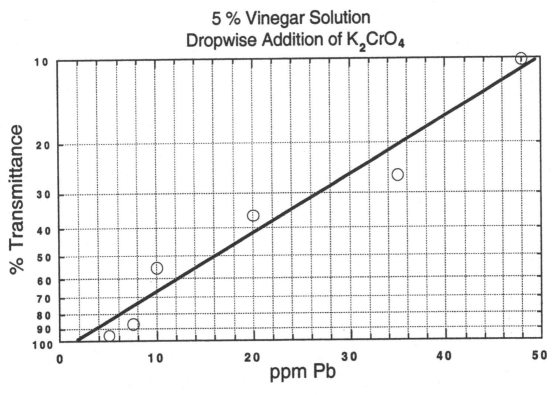

Experiment May Contain Lead

Date _____   Section number _____   Name _____

A. Gravimetric Method (Paint chip procedure)

1. Sample preparation

   (a) Nature, brand (if any) and source of sample _____.

   (b) Weight of original sample to nearest 0.01 g ...._____ g.

2. Observations upon:

   (a) Adding KI solution _____.

   (b) Adding $Pb(NO_3)_2$ if the results in 2(a) were negative _____.

● ●

3. Complete this part only if you performed the quantitative determination of lead:

   (a) Weight of cleaned and dried tube to nearest 0.001 g ...._____ g.

   (b) Weight of cleaned and dried tube + PbI_2 ...._____ g.

   (c) Calculations:

       (1) Net weight of lead iodide (PbI_2)
       (subtract line 3(a) from line 3(b)) ...._____ g PbI_2.

       (2) Weight of lead in this amount of lead iodide
       (multiply line 3(c)(1) by 0.45 ...._____ g Pb.

       (3) % lead in original sample
       (divide line 3(c)(2) by line 1(b) then multiply by 100 ...._____ % Pb.

● ●

4. What can you say regarding the safety of your sample in the use for which it is intended?

B. Spectrometric Method *(Vinegar leach procedure)*

1. Nature of earthenware utensil

   (a) General description _____.

   (b) Country of manufacture, place and date of purchase_____

   _____.

2. Observations after adding potassium chromate

   (a) To *blank* solution _____.

   (b) To *leach* solutions _____.

3. Spectrophotometer results

   (a) % transmittance of leach sample

   Tube #1 =   _____% T.

   Tube #2 =   _____% T.

   (b) Concentration in ppm lead in leach samples (from calibration graph)

   Tube #1 =   _____ppm Pb.

   Tube #2 =   _____ppm Pb.

   Average = _____ppm Pb.

4. Would your sample pass the FDA standards for lead safety?

Comments and conclusions on experiment

Experiment May Contain Lead

Date _____   **Section number** _____   **Name** _____

1. Is lead metal significantly soluble in water?

2. Are there lead compounds which will readily dissolve in water? If so, cite at least two examples. (You may wish to consult *The Merck Index* for help).

3. What is the difference between the "kinds" of lead referred to in Questions #1 and #2?

Think, Speculate, Reflect and Ponder

4. Which "kind" of lead referred to above in Questions #1 and #2 would you predict to be more toxic? Explain why.

5. List three uses for lead and in each case state why you think lead was chosen rather than some other substance (i.e., what properties does lead have that "endear" it to that particular use?).

(a)

(b)

(c)

6. Why was lead added to gasoline in the first place and then why was it removed?

7. What statement is the artist, Larry Hamill, trying to make in the sketch appearing at the beginning of this experiment?

Experiment 17

Caffeine Crystals
from Beverages

Sample From Home

Bring ONE of the following:

1. One hundred milliliters (3 $\frac{1}{2}$ oz) of a caffeine containing pop or soft drink, *or*

2. One hundred milliliters (3 $\frac{1}{2}$ oz) of a boiling water extract of coffee grounds previously made up to "drinking strength," *or*

3. One hundred milliliters (3 $\frac{1}{2}$ oz) of a boiling water extract of tea leaves previously made up to "drinking strength," *or*

4. An amount of solid instant tea/coffee sufficient to make about $\frac{1}{2}$ cup of drinking liquid (1 g coffee or $\frac{1}{2}$ g tea).

Pop and soft drink samples usually yield purer and whiter caffeine crystals; avoid espresso, latte and similar coffee drinks.

Objectives

The purpose of this experiment is to isolate crystalline caffeine and determine its approximate amount in some common beverages using an extraction technique. The use of a melting point as a means of identifying substances will also be illustrated.

Background

Caffeine and closely related compounds like theophylline (used by asthmatics) and theobromine (a compound containing *no* bromine!) are stimulants which are found in a wide variety of plants growing throughout the world. The most common sources are:

coffee (roasted ground seeds of the coffee shrub),

tea (dried leaves of various shrubs found especially in China, Japan, and India),

cocoa (roasted ground seeds of the small evergreen cacao tree found in tropical America),

maté (a tealike drink made from the leaves of a species of holly—the national drink in many South American countries) and

kola nuts (chestnut sized seeds of a tree indigenous to western tropical Africa, the West Indies and Brazil which are reportedly used in the Sudan both for chewing as well as a form of money).

It seems that whenever plants having a high caffeine content grow in a particular area, the native population uses extracts of the plant as a beverage.

The caffeine molecule.

One legend credits the discovery of coffee to a prior in an Arabian convent. Shepherds reported to the prior that goats who had eaten the berries of the coffee plant did not rest but gamboled and frisked about all through the night. The prior, mindful of the long nights of prayer which he had to endure, instructed the shepherds to pick the berries so that he might make a beverage from them. The success of his experiment is obvious today!

Although tea leaves contain considerably more caffeine than coffee grounds on a dry weight basis, a cup of coffee will actually contain more caffeine because more coffee than tea is used in making a cup of "drinking strength" brew. The several million pounds of natural (from decaffeinating coffee) and synthetic caffeine produced annually in the United States are used mainly in headache and "stay-awake" tablets. Because of its stimulant effect on the central nervous system, caffeine has also been used as an antidote to counteract the depressant effects of morphine poisoning. The lethal dose for caffeine in humans is estimated to be about 10 g (10,000 mg), but no deaths have been reported. This lethal dose would be equivalent to around 100 cups of coffee drunk all at once! (For a discussion of the caffeine content of specific brands of coffee, tea, and chocolate drinks, see these issues of Consumer Reports: September, 1976; October, 1979; May, 1985; September, 1987; October, 1994.) And for those other questions that you've always wanted to know about caffeine but were afraid to ask, you can now "Ask an expert" (see article on the next page).

The decaffeinating of coffee used to be done by extracting out the caffeine with liquid chlorinated hydrocarbons—a method similar to the extracting of caffeine from your beverage with methylene chloride in this experiment. Since such chlorinated compounds have been shown to cause liver damage and are implicated as carcinogens, a process using water extraction has now been developed and is widely used.

Ask an Expert: How Much Caffeine Is Too Much?

Question: What can you tell me about caffeine in the diet? How much is too much? Are the new extra caffeine soft drinks like "Jolt" damaging? I've read that caffeine has been found to speed weight loss by speeding up the body's metabolism. Is that true? Also, is there a difference in the caffeine in coffee, tea or soft drinks? I've also heard that caffeine has been linked to a high incidence of breast cysts in women.

Answer (from Maureen Raskin, GHC nutritionist and chronic disease epidemiologist): Caffeine is one of a group of compounds known as methyl-xanthines found in a wide variety of substances—from over-the-counter medications to foods where it occurs naturally. It may be present in headache, cold, allergy, menstrual pain and general pain-killing over-the-counter remedies as well as in diet pills, stay-awake pills and some prescription drugs. It is found naturally in coffee and tea, and in smaller amounts in chocolate and cocoa. The amount of caffeine in coffee and tea depends on how the beverage is prepared and which brands are used. I have included a summary table so you may make some comparisons.

Item	mg. of Caffeine
Coffee (5oz.)	
Drip	110-150
Percolated	64-124
Instant	40-108
Decaffeinated	2-5
Tea (5 oz.)	
1-minute brew	9-33
3-minute brew	20-46
5-minute brew	20-50
Chocolate Products	
Hot Coca (6 oz.)	2-8
Chocolate Milk (8 oz.)	2-7
Milk Chocolate (1 oz.)	1-15
Sweet Dark Chocolate (1 oz.)	5-35
Soft Drinks [†]	
Jolt	72
Mountain Dew	54
Tab	47
Coca-Cola	46
Shasta Cola	44
Dr. Pepper	40
Pepsi Cola	38
RC Cola	36

[†] Colas and pepper-type drinks derive less than 5% of their total caffeine content for the cola nut. The rest is added and typically is the caffeine obtained from raw coffee beans in the process of decaffeinating coffee.

Since there is no strong evidence that moderate amounts of caffeine are harmful to the *average healthy* adult, it is listed on the "Generally Regarded as Safe" [GRAS] list put out by the U.S. Food & Drug Administration. The Food Chemicals Codex of the U.S. National Academy of Sciences/National Research Council lists caffeine as a flavoring agent and stimulant.

Excessive consumption is considered to be more than 600 mg. per day (about 5-6 cups of strongly brewed coffee). However, individual tolerance to caffeine varies considerably. Persons who generally consume little or no caffeine show much greater sensitivity to it than those who consume caffeine on a regular basis. Symptoms of excess caffeine intake (known as caffeinism) can include anxiety, restlessness, delayed onset of sleep, headache, diarrhea and heart palpitations.

Caffeine exerts several effects on the body even when consumed in safe amounts. It stimulates the central nervous system, thereby producing the "lift" so many people seek. Depending on the dose, it can increase heartbeat, increase secretion of stomach acid, and yes, it can also increase one's resting metabolic rate. I do not recommend it for persons who are prone to stomach ulcers or for persons with diabetes since it tends to keep one's blood sugar elevated for a longer period of time after a meal. This last effect is why it is used in weight control products as a means to assist in reducing the feeling of hunger. Caffeine is used in many headache remedies, in combination with other drugs, because it constricts the swollen blood vessels in the head that cause headache and so helps speed relief.

Product	mg. per Tablet/Capsule
Vivarin	200
No-Doz	100
Dexatrim ¥	200
Diatac ¥	200
Exedrin	65
Vanquish	33
Anacin	32
Midol	32
Plain aspirin, any brand	0

¥ These products also come in caffeine-free

The time required for the body to eliminate caffeine varies from several hours to several days, depending upon age, if other medications are being used, and whether or not the individual smokes. The amount of time needed for the body to get rid of half the caffeine consumed (scientists call this the "half life" of a substance) is as follows: for children and smokers it is less than 3 hours, for the average nonsmoking adult, about 5 to 7 hours; for women taking oral contraceptives, up to 13 hours; for pregnant women on their last trimester, 18 to 20 hours; and for newborns (who do not have the enzymes needed to metabolize caffeine until several days after birth), 3 to 4 days. Caffeine crosses the placenta so pregnant women are giving it to the fetus, and it is secreted into human milk.

Several studies have been done which suggested an association between consumption of methylxanthines (caffeine and its close relative, theobromine) and fibrocystic breast disease (FBD). The amount of benefit gained from abstaining from caffeine remains unclear. While modest results were obtained from some studies, the largest study conducted by researchers from the National Cancer Institute failed to support such findings. Perhaps this is yet another example of the wide variation in individual response—limiting or avoiding caffeine may be helpful for some women yet of no value in reducing symptoms of FBD in other women. I guess like just about everything else, the word is "moderation."

Figure 17.1. Some sources of caffeine.

The procedure in this experiment calls for you to dissolve out (to *extract*) the caffeine from a water solution using the organic solvent methylene chloride, CH_2Cl_2. Caffeine is slightly soluble in water, but very soluble in methylene chloride. Thus mixing these two insoluble liquids together will cause most of the caffeine to dissolve preferentially in the organic methylene chloride layer. The sodium carbonate that is added reacts to form salts with many of the non-caffeine substances (especially tannins and tannic acid) which prevents them from also dissolving in the methylene chloride thus contaminating the caffeine.

The two liquid layers (water and methylene chloride) are then separated by, in this special case, a rather sneaky kind of filtration. Once wetted with water, the pores of a filter paper will only allow water to drain through them. When the two liquids are poured into this wet paper, the methylene chloride layer will stay trapped inside the filter cone while the water layer drains through. The product you get after evaporation of the liquid methylene chloride should consist of brown to snow white caffeine crystals weighing 15 – 30 mg.

The question then arises "How do you know that these crystals are indeed caffeine?" A quick and simple method for identification is a procedure called a *melting point determination*. Like boiling point and density, the melting point is also a physical constant characteristic for a particular substance. Just as all samples of solid water melt at 0 °C, all samples of pure caffeine melt at 238 °C. By comparing the observed melting point of an unknown compound with recorded melting points for known compounds, evidence for identity can be obtained. Your actual observed melting point will probably be a little below 238 °C, due largely to impurities present in your sample and the unsophisticated design of the melting point apparatus. Some coffees may give caffeine samples melting at 190 to 220 °C due to impurities, so do not be alarmed. Be aware that caffeine is a material which can *sublime* (go directly from a solid to a gas phase), so rapid heating of your sample is necessary so that you can see the solid melt before it "evaporates".

As we prepare to begin this experiment, you may find Table 17.1 both interesting and useful. The Austrian sport drink import *Red Bull* is currently leading in the battle of the caffeine wars, but at around $2 for a miserly 8.3 oz can, it is a pricey choice as well.

Red Bull (80 mg per 8.3 oz can)	115.5	Mountain Dew (0 in Canada)	55.0	Pepsi, Wild Cherry	38.0
Afri-cola	100.0	Mountain Dew, Diet	55.0	Pepsi, Diet	36.0
Java Water (125 mg per 16.9 oz)	88.8	Mountain Dew, Code Red	55.0	Pepsi, Wild Cherry Diet	36.0
Bawls Guarana (67 mg per 10 oz)	80.0	Kick Citrus	54.0	Aspen	36.0
Jolt	72.0	KMX	53.0	RC Cola, Diet	36.0
Krank 20 (100 mg per 16.9 oz)	71.0	Mello Yellow	51.0	Diet Rite	36.0
RC Edge	70.2	Mello Yellow, Diet	51.0	Coca-Cola, Classic	34.0
XTC Power Drink	70.0	Surge	52.5	Coca-Cola, Cherry	34.0
Java Water (125 mg per 16.9 oz)	88.8	Nehi Wild Red	50.1	Snapple Peach	31.5
Bawls Guarana (67 mg per 10 oz)	80.0	Tab	46.8	Snapple Raspberry	31.5
Jolt	72.0	Battery Energy Drink	46.7	Snapple Lemon	31.5
Krank 20 (100 mg per 16.9 oz)	71.0	Water Joe (60-70 mg per 16.9 oz)	46.2	Canada Dry Cola	30.0
RC Edge	70.2	Coca-Cola, Diet	45.0	A & W Creme Soda	29.0
XTC Power Drink	70.0	Shasta Cola	44.4	Nestea Sweet Iced Tea	26.5
Sun Drop, Diet	69.0	Shasta Cola, Cherry	44.4	Snapple Green Tea w/Lemon	24.0
Java Water (125 mg per 16.9 oz)	88.8	Shasta Cola, Diet	44.4	IBC Cherry Cola	23.0
Bawls Guarana (67 mg per 10 oz)	80.0	RC Cola	43.2	Barq's	22.5
Jolt	72.0	RC Cola, Cherry	43.2	A & W Diet Creme Soda	22.0
Krank 20 (100 mg per 16.9 oz)	71.0	Dr Nehi	42.0	Mistic Lemon Tea	18.0
RC Edge	70.2	Sunkist, Diet	42.0	Mistic Peach Tea	18.0
XTC Power Drink	70.0	Sunkist	41.0	Nestea Iced Tea	16.5
Sun Drop, Diet	69.0	Mr. Pibb	40.8	Cool (Nestea)	16.5
Aqua Blast (90 mg per 16.9 oz)	63.9	Mr. Pibb, Diet	40.5	Snapple Sweet Tea	12.0
Sun Drop, Cherry	64.0	Red Flash	40.5	Diet Cool (Nestea)	10.5
Sun Drop	63.0	Dr. Pepper	41.0	Snapple Sun Tea	7.5
Sugar-Free Mr. Pibb	58.8	Aqua Java (50-60 mg per 16.9 oz)	39.1	Snapple Sun Tea, Diet	7.5
Josta	58.0	Ruby Red Squirt	39.0	Canada Dry Diet Cola	1.2
Kick	57.6	Ruby Red Squirt, Diet	39.0	7 Up, Fresca & Sprite	0.0
Pepsi One	55.0	Pepsi	38.4	Hires and A&W Root Beer	0.0

Table 17.1. Caffeine content of beverages in milligrams per 12 oz of liquid.

Procedure

A. Preparation

Clean a 500 mL Erlenmeyer flask with soap and water. Record the kind of sample and, if it is an instant tea or coffee, record its weight. Prepare it for extraction by following ONE of the following three methods:

Pop: use 100 mL of a cola drink (check the ingredients on the label if in doubt). Pour all of it into the 500 mL Erlenmeyer flask and shake vigorously until most of the foaming due to carbon dioxide evolution ceases (about 5 minutes).

Instants: weigh out the amount of instant coffee (1 g) or tea (0.5 g) required for 1/2 cup and dissolve this solid in

Figure 17.2. Ready to extract caffeine using methylene chloride.

100 mL of cold water contained in the clean 500 mL Erlenmeyer flask.

Regular brews: bring to class 100 mL of tea or coffee extract previously made from tea bag, tea leaves, coffee grounds or the like. Pour this 100 mL into the clean 500 mL Erlenmeyer flask. If still hot, cool the brew to room temperature by adding 2 or 3 ice cubes to the flask contents.

B. Isolation

1. Dissolve roughly two grams of sodium carbonate [Na_2CO_3] in your sample in the 500 mL Erlenmeyer flask as prepared in one of the steps from Part A above.

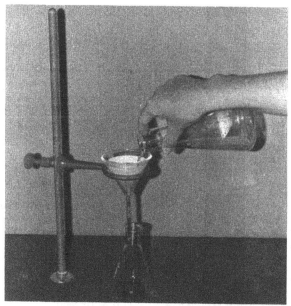

Figure 17.3. Pouring methylene chloride layer into wetted paper cone.

2. Add 25 mL of CH_2Cl_2 (methylene chloride) and agitate by "aggressively swirling" for 5–10 minutes; your lab instructor will demonstrate. DO NOT SHAKE or a persistent liquid/liquid *emulsion* will likely form.

3. Let the flask contents stand undisturbed for several minutes while you suspend a long stem funnel in an iron ring and place a collection beaker or flask underneath (size unimportant). Fit the funnel with a cone made from a 12 ½ cm filter paper and wet the paper thoroughly with water. See Experiment 3: *Recycling Aluminum Chemically* for the proper method of folding a filter paper if you have forgotten.

4. Slowly and carefully pour off and discard as much of the usually dark upper water layer as possible. (This is called *decantation*.) Leave behind in the flask all of the more *dense* methylene chloride lower organic layer together with a little remaining water. (See Experiment 2: *Going Metric* for a discussion of *density*.)

5. Slowly pour the entire contents of your 500 mL Erlenmeyer flask (the bottom layer of methylene chloride + upper layer of remaining water) into the previously wetted filter paper cone. Keep the liquid level up near the top of the paper cone until all of the flask contents have been thus transferred. You will see the upper water layer wick into the paper cone and drain through it. Left behind trapped in the filter paper will be the clear (usually) or maybe dark methylene chloride solution containing your

Figure 17.4. Water draining through paper leaving methylene chloride layer behind in paper cone.

dissolved caffeine. (If you have a froth/emulsion in your cone, you may carefully and gently stir it with a smooth end stirring rod to hasten the separation.)

6. While the last bit of water is draining through the filter paper cone, take a clean and bone dry 50 mL beaker and put your initials on it. (Use a pencil to write on the frosted glass circle on the side of the beaker.) Weigh this beaker to the nearest milligram (0.001 g) using a top loading electronic balance and record data on the report sheet. Don't forget to zero the balance.

 Your lab instructor can assist you in the use and care of these balances. (Be *warned* that these balances will quickly *spoil* you so that you will want to use nothing else!)

Figure 17.5. Weighing beaker on a top loading electronic balance.

7. Using a medicine dropper, transfer to the weighed beaker ONLY the clear or sometimes dark methylene chloride layer remaining in your filter cone. Take care not to suck up any frothy liquid or poke a hole in the bottom of the filter paper with your medicine dropper. (If desired, your lab instructor can help you clarify any brown coloration in the liquid resulting from coffee or tea samples.) If any water droplets are seen contaminating your methylene chloride solution, your lab instructor can show you how to remove them also using anhydrous sodium sulfate, Na_2SO_4.

8. Go to the HOOD and place your beaker on a hot plate, but DO NOT LEAVE BEAKER UNATTENDED. Avoid breathing the methylene chloride vapors. Watch the liquid contents closely with an eagle's eye and, when the liquid level in the beaker gets down to about the thickness of a penny,

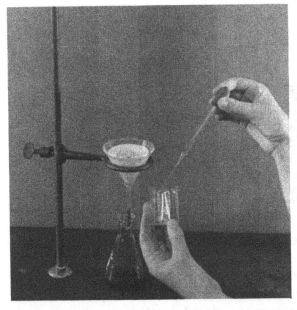

Figure 17.6. Transferring methylene chloride solution containing caffeine into a beaker.

Figure 17.7. Evaporating methylene chloride solution on a hot plate in a hood.

IMMEDIATELY REMOVE the beaker from the hot plate and set it aside in the hood to cool while the small remainder of methylene chloride liquid evaporates (under 5 min). Methylene chloride boils at only 40 °C, so you can safely grab the beaker with your fingers—a better option than unnecessarily trying to hold the beaker with tongs.

Observe and record your observations on the report sheet; also make a sketch of the solid residue in the beaker with the aid of a microscope.

Reweigh the beaker to the nearest 0.001 g. (Be sure to use the same balance that you used before.) Now we are ready to try to find out if what we are looking at is really caffeine.

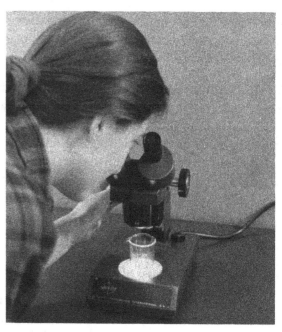

Figure 17.8. Examining caffeine crystals in beaker using a microscope.

C. Identification (melting point determination)

Note. You only get one chance to take a melting point, so read this section through carefully before beginning. Examine the melting point apparatus in both the diagram below in Figure 17.9 and the photograph on the next page (Figure 17.11). Be sure to use a cork (*not* a rubber stopper) having a hole that snugly holds a thermometer reading 250 °C or higher. Clamp the 8 inch test tube horizontally about six inches above your desk top and insert the cork + thermometer to make sure everything fits properly.

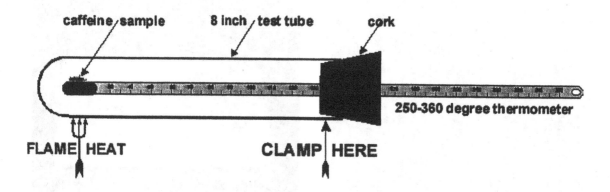

Figure 17.9. Apparatus for a melting point determination.

Using a spatula, carefully transfer the pile of caffeine onto the top of the thermometer bulb and gently reinsert the thermometer into the test tube. Be careful to keep the caffeine sample on top of the bulb. See Figures 17.9 and 17.10. If you have both thought *and* read ahead, the numbered scale on the thermometer will be facing you for easy reading.

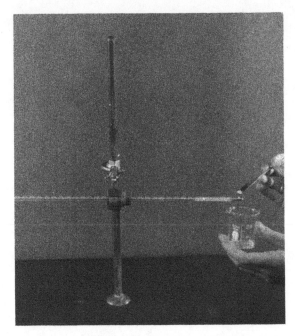

Figure 17.10. Transferring caffeine crystals to a thermometer bulb.

Adjust a lit Bunsen burner to give a cooler flame by cutting down somewhat on the air supply (remember Experiment 1: *The Ubiquitous Bunsen Burner*). No inner hot cone should be present, but neither should you see any yellow in the flame. Your lab instructor can assist with this adjustment if you have trouble. Proceed to heat the bottom of the test tube with the flame, moving it back and forth directly under the thermometer bulb.

Watch the caffeine crystals closely. As you near the melting point, the crystals will first collapse to form an opaque semisolid mass. Record as the actual melting point the temperature at which this mass suddenly turns water-clear or tannish-clear. Let your apparatus cool; then disassemble it and clean the thermometer bulb.

With the aid of a microscope, observe and comment on the white "fog" coating the inside of your 8 inch test tube. (Hint: place the "fog" side of the test tube up under the microscope. Touch a pencil tip to the top of the tube and focus on the pencil tip to locate the proper focus height for the scope to see the "fog.")

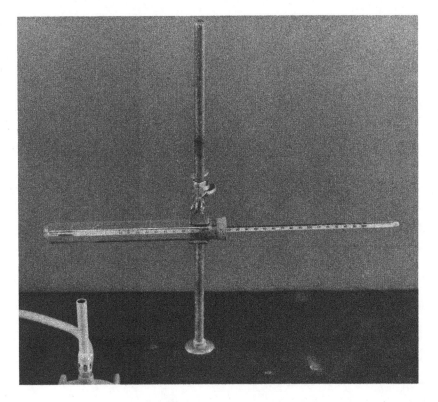

Figure 17.11 Ready to run a melting point determination.

Caffeine Crystals

Date _____ **Section** _____ **Name**_____

A. Preparation

Nature of sample (beverage type, brand) _____.

Amount of sample taken _____.

B. Isolation

Appearance of solid residue in beaker after evaporation of methylene chloride

Sketch of caffeine crystals

Weight of beaker + dry residue of caffeine .... _____grams.

Weight of beaker (empty and dry) .... _____grams.

Total weight of residue (caffeine) .... _____grams.

C. Identification

Melting point of residue .... _____°C.

Melting point of pure caffeine (consult *The Merck Index.*) .... _____°C.

Description and identity of coating on 8 inch test tube

_____.

D. Comments and conclusions on experiment

Caffeine Crystals

Date _____   **Section** _____   **Name**_____

1. In the caffeine isolation steps:

   (a) Which layer in the actual extraction step was the methylene chloride—the top or the bottom?

   (b) What physical property of methylene chloride would cause it to be the particular layer that you said it was in (a) above?

2. A thin film of white solid can usually be seen on the cooler parts of the test tube and on the thermometer inside the apparatus after completing the melting point determination. What is this solid and how did it get there?

3. Define the term sublimation.

4. At what temperature does caffeine start to sublime rapidly? (HINT: check *The Merck Index* or background discussion to this experiment.)

5. Give the brand name and ingredient list of two drugs that contain caffeine.

   (a)

   (b)

Think, Speculate, Reflect and Ponder

6. The symptoms of withdrawal from an addictive drug are irritability, desire for the addicting drug, and physical discomfort. All of these symptoms are relieved by a dose of the addicting drug. Using this as an indicator, is caffeine an addicting drug?

7. Do other extremely widely used drugs like alcohol and the nicotine in cigarettes fit the profile of addictive drugs? State one reason why you would (or would not) favor a law making the sale of one or both of these drugs illegal.

8. In April, 1994, the top executives from the seven largest tobacco firms swore before the U.S. House of Representative that nicotine is not an addictive drug. Explain their reasoning and why they would take such a position.

Experiment 18

Fluoridation
(What! Another Plot to Poison Us?)

Sample From Home

Bring a small (5 mL) sample of clear water from the tap, well, stream, lake, ocean etc.

Objectives

In this experiment you will determine the fluoride content of a water sample using a standardized procedure. You will be able to compare your results with the fluoride limits set by your local public health department.

Background

". . . The main communist plots by the internal traitors in the U.S. are fluoridation, disarmament and federal aid to education. . . This (fluoridation) will positively degenerate Christian Americans by dulling their brain and crippling their bodies. Fluoridation causes an increase of 30% to 125% in anemia, diabetes, heart disease, stroke and cancer: fluoridation also causes abortions and Mongoloid births. Do you want your children to become progressively dull and morons? Do you want your grandchildren to be born Mongoloids and cripples? Then get out and fight fluoridation day and night."

Such ringing rhetoric rode the crest of a wave of anti-intellectualism in the scare campaign against fluoridation in the mid 1960s, receiving wide support from such well known organizations as the Ku Klux Klan and the John Birch Society. Ever since 1941, when a group of dentists in Wisconsin proposed the fluoridation of public water supplies to reduce tooth decay, the battle was joined and still continues today.

As approved and recommended by the United States Public Health Service, water fluoridation involves adding fluoride to the water supply until the concentration level falls within limits set in Table 18.1. The permitted amounts are tied to temperature because of its effect upon water consumption.

Annual average of maximum daily air temperatures (°F)	Recommended control limits for fluoride concentrations (ppm)		
	LOWER	OPTIMUM	UPPER
50.0–53.7	0.9	1.2	1.7
53.8–58.3	0.8	1.1	1.5
58.4–63.8	0.8	1.0	1.3
63.9–70.6	0.7	0.9	1.2
70.7–79.2	0.7	0.8	1.0
79.3–90.5	0.6	0.7	0.8

Table 18.1 Official Control Limits for Fluoridation.

The actual fluoridation chemicals used, in order of decreasing cost, are liquid fluorosilicic acid, sodium fluoride, ammonium silicofluoride, sodium silicofluoride, and fluorspar. Fluoridation costs per person in the United States are typically around 15 – 20 cents per year. The addition of the controlled amounts of fluoride indicated in this table has been shown to reduce tooth decay 60–65% at age 15 if children consume fluoridated water from time of birth. Continued inhibition carries on into adult life.

Strangely, persons who are against fluoridation often seem willing to accept the necessity for chlorination of water supplies, in spite of the fact that the free elemental chlorine gas used is not found naturally in any substance on earth. Neither are some of the reaction products of chlorine with organic material in water found naturally in any water supply; some of these reaction products include carbon tetrachloride and chloroform, known possible carcinogens. Check the difference between chlorine/chloride and fluorine/fluoride in your chemistry text. (The ozonolysis of water has none of these problems of chlorination; maybe the United States will eventually follow the lead of Europe and move towards the use of ozone to purify water supplies.)

Fluorides, on the other hand, are one of many trace ions naturally present in most water supplies and foods which, like many other elements, actually appear to be necessary in small amounts for good health — much like iodide that is required to prevent the development of the condition known as goiter. Table 18.2 on the next page lists some typical natural fluoride concentrations in foods. (Since most fluorine in food is organically bound, only about 50% may actually be taken in by the body as fluoride ions.)

Interestingly, tea is exceptional in its ability to concentrate fluoride, and some special teas have been reported to contain as much as 1530 ppm (parts per million) fluoride on a dry basis. In fact, tea is a major source of fluoride in the tea-drinking New Zealanders' diet because their drinking water is practically free of this element.

Food	Fluorine (ppm)	Food	Fluorine (ppm)
FLUORINE REPORTED IN FOOD AS CONSUMED			
Milk	0.07–0.22	Pork chop	1.00
Egg white	0.00–0.60	Frankfurters	1.70
Egg yolk	0.40–2.00	Round steak	1.30
Butter	1.5	Oysters	1.50
Cheese	1.6	Herring (smoked)	3.50
Beef	<0.20	Canned shrimp	4.40
Liver	1.50–1.60	Canned sardines	7.30–12.50
Veal	0.20	Canned salmon	8.50–9.00
Mutton	<0.20	Fresh fish	1.60–7.00
Chicken	1.40	Canned mackerel	26.89
Pork	<0.20		
FLUORINE REPORTED IN DRY SUBSTANCE OF FOOD			
Rice	<1.00	Honey	1.00
Corn	<1.00	Cocoa	0.500–2.00
Corn (canned)	<0.20	Milk Chocolate	0.500–2.00
Oats	1.30	Chocolate (plain)	0.50
Crushed oats	<0.20	Tea (various brands)	30.00–60.00
Dried beans	0.20	Cabbage	0.31–0.50
Whole buckwheat	1.70	Lettuce	0.60–0.80
Wheat bran	<1.00	Spinach	1.00
Whole wheat flour	1.30	Tomatoes	0.60–0.90
Biscuit flour	0.00	Turnips	<0.20
Flour	1.10–1.20	Carrots	<0.20
White bread	1.00	Potato (white)	<0.20
Ginger biscuits	2.00	Potato (sweet)	<0.20
Rye bread	5.30	Apples	0.80
Gelatin	0.00	Pineapple (canned)	0.00
Dextrose	0.50	Orange	0.22

Table 18.2 Fluorine Content of Foods.

In many parts of the world, the natural fluoride concentration is well in excess of 1.0 ppm. The largest area of high fluoride concentration in the United States is in the panhandle/West Texas region where such concentrations average 3–6 ppm; they have stood at 8.0 ppm in Bartlett, Texas since at least as far back as 1901. Studies have shown that similar population groups in other countries, including the former Soviet Union, have been drinking high concentrations of fluoride for generations. These high fluoride levels had no adverse effect on the health of the affected population when compared to a control population sample nearby. This control population was subjected to similar environmental stresses except for having a very low fluoride level in its drinking water (see tables on the next page). The main undesirable effect — and one which is consistently noted with high (over 2 ppm) fluoride content of drinking water — is principally a cosmetic one due to mottling of the teeth to give a brown stained appearance.

Tables 18.3 and 18.4 below graphically illustrate the effects of fluoridated water upon teeth. The drinking water in Colorado Springs has contained 2.0–2.5 ppm fluoride since at least the turn of this century, whereas the concentration in Boulder, Colorado, water (about 100 miles away) is only 0.0–0.1 ppm.

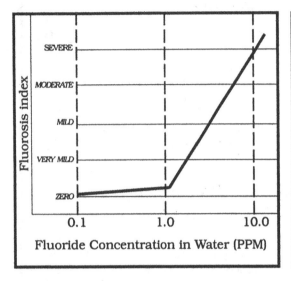

Table 18.3 Degree of Mottled Enamel and Fluoride Concentration in Water.

Table 18.4 Tooth Mortality in Natives of Boulder and Colorado Springs (adjusted for number of school years completed).

Threshold levels for optimum tooth protection begin around 0.7 ppm fluoride. Sea water ranges from 1.0–1.4 ppm fluoride and almost all the world's fresh water drinking supplies contain over 0.05 ppm fluoride, largely due to the leaching of this element from certain rocks such as calcium fluoride (commonly called fluorspar or fluorite). Table 18.5 shows the distribution of naturally occurring fluoride in water over 0.7 ppm throughout the United States.

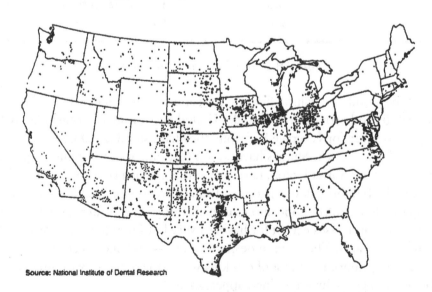

Table 18.5. Distribution of water supplies in the U.S. containing over 0.7 ppm natural fluoride.

As necessary, water supplies may have fluoride added to bring up the concentration to optimum levels (near 1.0 ppm). By the year 2000, 162 million persons (66%) of the U.S. population was drinking fluoridated water, an increase of 3.7% over 1992 figures (Table 18.6).

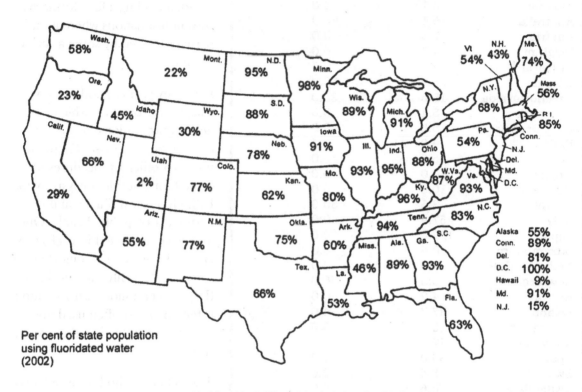

Per cent of state population
using fluoridated water
(2002)

Table 18.6 Fluoridation in the United States.

Fluoridation of public drinking water was first begun in the United States in 1945. Since then, long term studies on large populations have convinced most authorities in the safety and desirability of fluoridation for all except those on kidney dialysis. However, an increasing number of professionals are now feeling that perhaps continued fluoridation, at least in the United States, is unnecessary due to the availability and widespread use of fluoridated toothpastes. Serious questions are also currently being raised about a possible damaging role of fluoride in certain bone disease.

But the following quote from a letter to a professional chemistry news magazine does much to put the debate in perspective:

"I watched, chagrined, as my husband, who is a Ph.D. chemist, lavished fluoride toothpaste on my son's toothbrush and my son gleefully swallowed every bit of it. I have only a M.S. degree in chemistry, but I managed to convince my husband to skimp on the toothpaste until I could obtain a tube of ingestible toothpaste containing no fluoride. I am less worried about water fluoridated at 1 ppm than I am about young children swallowing toothpaste with fluoride at levels of 1000 ppm."

COUNTRY	POPULATION (MILLIONS)	PERCENT OF POPULATION DRINKING ARTIFICIALLY FLUORIDATED WATER
Albania	3.1	0.0
Australia	16.1	66.0
Austria	7.6	0.0
Belgium	9.9	0.0
Bulgaria [a]	9.0	0.0
Canada	25.9	50.0
Czechoslovakia	15.6	20.0
Denmark	5.1	0.0
East Germany [b]	16.6	9.0
Finland [c]	4.8	1.5
France	55.6	0.0
Greece	10.0	0.0
Hungary	10.6	0.0
Ireland	3.5	50.0
Italy	57.4	0.0
Japan	122.0	0.0
Luxembourg	0.4	0.0
Netherlands [d]	14.6	0.0
New Zealand	3.3	66.0
Norway	4.2	0.0
Poland	37.7	4.0
Portugal [a]	10.3	0.0
Romania	22.9	0.0
Spain	39.0	<1.0
Sweden	8.4	0.0
Switzerland	6.6	4.0
United Kingdom	56.8	9.0
United States	243.8	50.0
U.S.S.R. [e]	284.0	15.0
West Germany [b][f]	61.0	0.0
Yugoslavia	23.4	0.0

(a) One experimental treatment plant now discontinued
(b) Before German reunification in 1990
(c) One small experimental treatment plant
(d) Discontinued in 1976 after 23 years of experiments
(e) Prior to Commonwealth formation in December 1991
(f) Discontinued in 1978 after 18 years of experiments

Table 18.7 Fluoridation Worldwide.

Most major developed countries do not fluoridate their water supplies. Table 18.7 shows the fluoridation trends in different countries of the world, including a few (formerly) communist nations where it was perhaps termed by some a "capitalist plot."

For many years a good sensitive analytical procedure which could detect fluoride ions in low concentrations did not exist. There are now two official procedures accepted for fluoride analysis by the U.S. Geological Survey and departments of public health. One uses what are called ion selective electrodes which can quickly measure electronically the fluoride ion $[F^{-1}]$ concentration. This method is the one most often used now, but the apparatus is rather expensive.

The other method — an adaptation of which this experiment follows — was developed in 1954 and determines the amount of fluoride ion by mixing the sample with a red-colored complex ion formed by reaction of an organic dye (*eriochrome cyanine R*) with *zirconium* ions. (This constitutes the mixed indicator solution used in this experiment).

This rather exotic red *zirconium/dye* complex ion is chosen because fluoride ions react with it to displace some of the zirconium ions from the organic complex which causes a loss of some of the red color. Thus the *more red* the complex remains after the water sample is added to it, the *less fluoride* is present; conversely, the *less red* it becomes, the *more fluoride* there is present. (Tin chloride is added to inactivate any chromate, residual chlorine, or other similar oxidizing agents which, if present, would interfere with the analysis.)

The intensity of this red color will be measured electronically with a Spectronic 20 spectrophotometer, whose principle and method of operation were described in Experiment 16: *Warning: This Experiment May Contain Lead.* Comparison of the observed light absorbance of your sample with a calibration curve will permit you to read directly the concentration of fluoride in your unknown water sample. Although absorbance numbers are harder to read off the Spectronic 20 than per cent transmittance, absorbance has the advantage of being directly proportional to color intensity and hence fluoride concentration. This means that a plot of *fluoride ion concentration* against *absorbance* will give a straight line graph.

Procedure

Although this analysis involves very few steps, a high degree of meticulousness will be required of you for accurate results. In order to check your precision and maximize the validity of your results, this procedure has you run your unknown water sample in duplicate.

1. Scrupulously clean two 4 inch test tubes with soap and water and rinse thoroughly with distilled water. Dry by rinsing twice with a little acetone (CAUTION: flammable liquid. Make sure no flames are near.) and then, while holding with your wire test tube holder, passing through a Bunsen burner flame until completely dry. Obtain a 2 mL pipette, suck some distilled water up into it using a pipetter bulb (Experiment 13: *Vitamin C*) and allow the liquid to drain back out. Your lab instructor can assist in this operation. If the pipette is clean, no water droplets will remain clinging to the inside of the pipette. The presence of such droplets means that the pipette is dirty (however clean it may look!). Take the easy way out; return the pipette and obtain a (hopefully) clean replacement.

*Who cares about fluoride?
I just want my 2.1 million
gallons of water each year*

2. Now that your necessary equipment is clean, you can proceed to test for traces of fluoride ions in your water sample from home. Your water sample must be absolutely clean and free of sediment; if not, see your lab instructor. Flush out the 2 mL pipette with your water sample and transfer exactly 2.00 mL of unknown water to each of your two clean and dry 4 inch test tubes. Using the medicine dropper provided in the reagent bottle, add exactly one drop of stannous (tin) chloride solution to each of the two samples. Let the drop fall from the dropper tip with a steady hand. Mix the contents using a rapid side-to-side agitation by bouncing the tube back and forth between your fingers. This can best be done with one hand, but needs to be demonstrated by your lab instructor. Then let the test tubes stand for ten minutes.

Note. You will be told whether your busy lab instructor will elect to dispense the mixed indicator reagent directly into your tubes instead of you doing this yourself, as described in the next step 3 which follows.

3. Go to where the **mixed indicator reagent** is located and, using the burette or pipette set up for this purpose, add exactly 5 mL of this reagent to each of your two samples. The pipette uses the same automatic pipetter bulb seen in Experiment 13:

Vitamin C in Your Diet. Your lab instructor can again assist in showing you how it works.

After the mixed indicator reagent has been added, place a small piece of plastic wrap over the mouth of each tube, mix well by inverting the contents several times, and allow the test tubes to stand for a minimum of 30 minutes.

4. Carefully *scrutinize* the light red-orange solutions in your tubes by holding them up to the light. A better way is to look down on top of the solutions and thence down through the whole depth of solution in the tube against a black back-

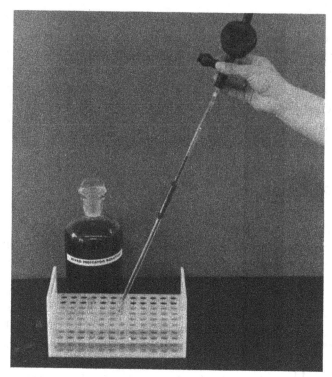

Figure 18.1. Adding mixed indicator reagent to water sample using a pipette.

ground (as described in Experiment 16: *Warning: This Lab May Contain Lead*). **In the unlikely event that *any* haze or cloudiness is discernible, you must proceed to Step 5.** This sediment haze needs to be removed from the solution by centrifugation before measuring the absorbance of your liquid with the spectrophotometer. Water samples low in sulfate ion concentration will probably not give any cloudiness, in which case you can *proceed directly to Step 6.*

Note the presence or absence of any such haziness on the report sheet.

5. PERFORM THIS STEP ONLY IF YOUR SAMPLES ARE CLOUDY.

If centrifuges are available (Figure 18.2 on the next page) which can accommodate the 4 inch tubes, spin them for 5 minutes. (Your lab instructor will demonstrate). If only the smaller 3 inch capacity centrifuges are available, proceed as follows. First clean and dry four 3 inch test tubes in the same manner as you did the 4 inch tubes back in Step 1. Number them in pairs, for instance AA and BB, and transfer your two red-orange solutions into these two pairs by pouring the solution from each 4 inch test tube into the two 3 inch test tubes. Divide up each sample as equally as possible between each pair of two 3 inch tubes. Place the two pairs (4 tubes total) into a centrifuge so the pairs are positioned opposite each other and spin them for at least five minutes.

Obtain two *cuvette tubes* (precision manufactured test tubes) rinsed in distilled water but not necessarily dry. Be careful with them as they are expensive. Using a clean, dry capillary pipette (long nose medicine dropper), carefully suck up some of the clear liquid from one four inch tube (or one pair of 3 inch centrifuge tubes) and squirt about 1 mL (20 drops) into a cuvette tube. Agitate the liquid around in the tube and discard. (This serves to wash the tube free of distilled water.) Then fill the cuvette about two-thirds full with the remainder of the now clear liquid. Be careful not to suck up any sediment. If you accidentally stir up the sediment on the bottom of a centrifuge tube, spin it down again for another 5 minutes.

Repeat this transfer technique with your duplicate sample in the other 4 inch tube (or pair of 3 inch centrifuge tubes). You are now ready to use the spectrophotometer. *Proceed directly to Step 7.*

Figure 18.2. Two typical centrifuges.

6. Obtain two *cuvette tubes* (precision manufactured test tubes) rinsed with distilled water but not necessarily dry. Be careful with them as they are expensive. Using a clean and dry capillary pipette (long nose medicine dropper), squirt about l mL (20 drops) from one of your 4 inch test tubes from Step 4 into a cuvette tube. Agitate the liquid around in the tube and discard the liquid. (This serves to wash the tube free of distilled water.) Then fill the cuvette about two-thirds full with the rest of the solution left in the 4 inch test tube.

Repeat this transfer technique with your duplicate sample in the other 4 inch test tube. You are now ready to use the *spectrophotometer* described in the next step and shown in Figure 18.3 on the next page.

7. Refer to Experiment 16: *Warning: This Lab May Contain Lead,* for a description of the operation of the Spectronic 20. Set the wavelength dial at 540 mμ (540 nm). After adjusting the zero control knob, insert the provided cuvette containing the *reference blank solution* (2 mL of distilled water + 1 drop stannous chloride solution + 5 mL mixed indicator solution) into the machine and set the needle to read 0.500 *absorbance*. Remember to handle the cuvette tubes by the top and to wipe them clean and dry on the outside with tissue before inserting them into the instrument.

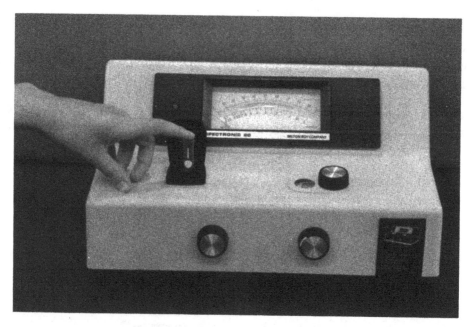

Figure 18.3. Spectrophotometer for measuring the light absorbance of solutions.

Insert one of the two cuvettes containing your duplicate water samples into the spectrophotometer and record the absorbance on the report sheet. Repeat with the other duplicate water sample. Remember to align the vertical mark at the top of the tube so that it faces you and close the instrument cuvette cover cap before taking your reading. Examine the lower absorbance scale carefully to be certain you read it correctly (your lab instructor can assist you here). Note that the value of each division changes in different parts of the scale. (Absorption is actually a logarithmic scale.) When you are finished, wash out the two cuvettes containing your samples thoroughly with distilled water and place them into a cuvette rack in an inverted position.

Reasonably precise absorbance readings should lie within 0.01 absorbance units of each other. If your two values do not deviate from each other by much more than this, your experimental technique was done carefully so congratulate yourself. Read off the actual fluoride ion concentration of your unknown water sample to the nearest 0.1 ppm from a calibration curve supplied by your lab instructor and compare it with values, if any, obtained from your public health department. (A brief discussion of how a spectrophotometer works and how to read a graph may be found at the end of the background section to Experiment 16: *Warning: This Experiment May Contain Lead.*)

This procedure will accurately handle fluoride ion concentrations up to around 3 ppm. Observed absorbances of less than 0.10 indicate that you likely have a sample containing more than 3 ppm fluoride. In such cases, the best accuracy can be obtained either by making a prior dilution of your water sample (necessitating repeating the experiment) or changing the absorbance setting of the reference blank solution to a higher value (0.600 or 0.700 absorbance) and running a new calibration curve. See your lab instructor for help if this becomes necessary.

Fluoridation

Date _____   **Section** _____   **Name**_____

A. Water Sample
   (a) Nature and source of water sample _____

   (b) Geographical location where sample taken _____

   (c) Water district (if tap water used) _____

B. Sample Preparation (steps 1–6)
   Appearance of solutions after adding mixed indicator reagent (step 4)

C. Spectrophotometer
   (a) Data (step 7)

Tube	Absorbance Readings	ppm Fluoride Concentration (from curve)
1		
2		
3 (optional)		
reference blank	0.5	0.0

   (b) Average of fluoride concentrations above _____

   (c) Reported fluoride level
         (Check with water department or public health department.) _____

   (d) Recommended optimum fluoride concentration (see Table 1) _____

D. Comments and Conclusions on experiment

Fluoridation

Date _____   Section _____   Name_____

1. Data handling

   (a) What would be the difference between the precision of your results and their accuracy?

   (b) Comment on the precision of your data. (Check the last page of the procedure for this experiment.)

2. Bottles of sodium fluoride have a skull and crossbones emblem on them to warn that the substance is a dangerous poison; it is sometimes used as a rodenticide and wood preservative. However, sodium fluoride may also be added to public drinking water. Explain.

3. Name two common everyday man-made materials that most people use which contain fluorine. (Consult the index in a chemistry text, if necessary, for help.)

Think, Speculate, Reflect and Ponder

4. Foods contain much of their fluorine in combination with carbon compounds. Explain how this "organic" fluorine differs from ionic fluorine like that in sodium fluoride. (Hint: What type of bond would hold fluorine to carbon atoms?)

Experiment 19

Sunglasses, Ultraviolet Radiation, and You

Sample From Home

Bring at least one pair of sunglasses to lab for this experiment. Get as much information about that pair of glasses as possible. The cost, manufacturer, and any ultraviolet and infrared light blocking claims by the manufacturer are all relevant to this experiment. Look for tags like "blue blockers" or "UV protection." You may be able to find information about glasses you already own on the Web.

Objectives

The purpose of this experiment is to allow you to measure the light blocking ability of your sunglasses. To do this you will watch the lab instructor operate a scanning spectrophotometer. The data that you will collect for each pair of glasses will give you information about their ultraviolet, infrared, and visible light transmission (or absorption) characteristics and even allow you to speculate about the visible spectrum of any colored lens.

Background

The radiation bathing the earth from our sun has a very wide spectral range. This means that the incoming light is made up of many different wavelengths (colors). Included in this life-giving radiation, however, are wavelengths of light that are harmful to organisms on the surface of the earth. Specifically, light in the ultraviolet region of the solar spectrum can be damaging to many organisms including human beings. Fortunately, our environment provides protection for the organisms that might be exposed to this harmful radiation.

The atmosphere above the earth acts as a radiation filter which removes significant amounts of ultraviolet radiation *before it strikes the earth*. The important chemical member of this atmospheric filter is *ozone* (O_3). Ozone is involved in a continuous cycle of creation and destruction in the upper atmosphere called the Chapman Cycle (or Chapman Mechanism) that also involves O_2 molecules (ordinary oxygen) and O atoms. The net result of the Chapman Cycle is the transformation of incoming energy in the ultraviolet wavelength into energy in the infrared wavelength. This prevents most of the ultraviolet radiation from our sun from reaching the surface of the earth. This process occurs far above the surface of the earth and at a safe distance above plants and animals that can be damaged by ultraviolet radiation. It also acts to heat the upper regions of the stratosphere causing a natural temperature inversion there, but that is another story (See Experiment 20: The Apparent Molecular Weight of Air). Let's concentrate on the ozone layer first.

The protective layer of ozone has been present in our upper atmosphere for millions of years. It is maintained, like many other atmospheric chemical cycles, by dynamic processes that involve both gases already in the upper atmosphere and those diffusing upward from the surface of the earth. In the 1930s, scientists at the Du Pont Company synthesized a family of chemicals called chlorofluoro-carbons (CFCs). These inert molecules were synthesized to be used in refrigeration as a replacement for the toxic chemicals then in use such as ammonia and sulfur dioxide. Due to their inertness, CFCs do not react to any appreciable degree with light or with chemicals that they encounter in the lower atmosphere (the troposphere, where we release them, as you will soon learn). They were very well designed for their job as refrigerants and in fact can be breathed by human beings with no danger. (Compare this to ammonia or sulfur dioxide, which are both toxic and stinkers besides!) After the Second World War, this property of chemical inertness made this class of compounds very useful in a variety of industrial applications beyond refrigeration such as cleaning agents and foam blowing agents, and therefore they provided our society with many benefits while greatly expanding their use even further than before.

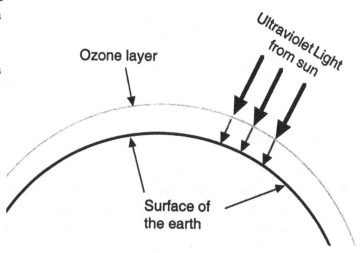

Figure 19.1. Simplified representation of UV filtering by the earth's ozone layer.

Since their introduction, these totally synthetic molecules have been escaping from their containers. They get loose in the lower atmosphere due to human beings; we release them on purpose or by accident from our air conditioners, refrigerators, solvent baths, etc. Since they don't react in the lower atmosphere, they become uniformly mixed in the lower atmosphere and slowly diffuse upward through the atmosphere. There is no doubt about their presence in the stratosphere: we have been able to analytically detect them there since the 1970s. Their presence there is a scientific certainty—radio talk show comments aside.

The average transit time for these compounds from the surface of the earth to the upper atmosphere has been estimated to be between one and two decades. This means that the chlorofluorocarbons released in the late 1960s and 1970s are just now arriving and causing problems in the ozone layer. Causing problems? But aren't they inert? Read on.

When CFCs finally arrive in the upper atmosphere (in a region called the stratosphere), their inertness becomes a thing of the past. The incoming solar radiation at that altitude is of sufficient energy (see Table 19.1) to break apart these molecules and free their previously strongly bound chlorine atoms (all of the original *chloro*fluorocarbons had some chlorine atoms). The newly freed chlorine atoms are the real culprits in the destruction of ozone in the stratosphere. Each chlorine atom takes part in a cycle of ozone destruction that destroys tens of thousands of ozone molecules before that particular chlorine atom is removed or converted to an atmospherically inert (chlorine reservoir) species. This means that the chlorine atoms contributed to the stratosphere from CFCs are very effective catalysts for ozone destruction. But what happens as the ozone layer thins?

The amount of ultraviolet (UV) radiation that strikes the earth under the ozone layer has *never* been zero since the ozone layer is not completely opaque to UV rays; some UV has always made it through. And, indeed, some UV exposure is required to initiate the production of vitamin D in human beings and probably also plays a role in genetic

Wavelength (nm)	Spectral Region	Relative Energy
10 to 380	Ultraviolet	Highest
380 to 750	Visible	In between
750 to 30,000	Infrared	Lowest

Table 19.1 Data for light of different wavelengths.

changes (on an evolutionary time scale); however, as the ability of the ozone layer to block UV decreases (because of ozone depletion due to CFCs), the amount of UV radiation striking the earth is increasing. The effects of increased surface UV radiation upon human beings is well documented: more UV means more skin cancers and more eye cataracts for human beings. The effects on crops, ocean algae, and animals is not as well understood but may also be quite damaging.

The Montreal Protocol—an international environmental agreement initially penned in 1987—and subsequent amendments (1990, 1992, 1995, 1997, and 1999) has led to the complete ban on the CFCs most hazardous to the health of stratospheric ozone. These chemicals are being supplanted by replacements that have much shorter atmospheric lifetimes. The fact that 46 countries signed the original planet-wide treaty and this number has risen to 175 countries shows the internationally agreed upon importance of ozone depletion.

The prognosis for stratospheric ozone has, therefore gotten rosier, and because of the decision to limit CFC release, stratospheric ozone will probably return to pre-1980 levels by approximately 2050. But until then, stratospheric ozone will still be less than it would be naturally—that is without human CFC release—especially over the poles, and humans' exposure to UV radiation will also be higher that before. Recent estimates are 6-14% higher UV in the mid latitudes than the early 1980s.

Because of the increased danger of UV radiation even at mid latitudes (where most people in the northern hemisphere live) we will have to take more stringent precautions against this danger. In the past we have protected our eyes from dangerous UV radiation by the use of UV blocking sunglasses, and we have protected our skin by using UV opaque sunscreen lotions; both contain molecules that do not transmit UV radiation. Sunglasses are designed to selectively block some wavelengths of light while allowing other wavelengths to pass through the lens and into your eye. Since there is no informational benefit in allowing UV into your eyes (since UV is invisible to our eyes and therefore can not help you *see* something), it is desirable that all ultraviolet radiation be blocked by sunglasses. Visible colors, however, can be selectively filtered by sunglasses to benefit the wearers in the particular light situation that they are in. For instance, dark mirrored sunglasses help the midday, full-sun skier who is skiing at high altitudes. Relatively light-colored, yellow sunglasses aid a skier in fog or blizzard conditions. Good skiers carry both in case the conditions change quickly. Getting caught in very bright light conditions when skiing (in which light comes from the sky and up off the snow) can lead to a dangerous condition called snow blindness.

Procedure

This experiment is designed to allow you to measure the light filtering ability of a pair of your own sunglasses. To accomplish this you will use (actually, watch your lab instructor operate) an instrument that measures both the wavelength and intensity of light that passes through your sunglasses. This instrument sends a beam of light through the sample (your sunglasses) and measures what comes through from the longest visible light wavelengths (about 780 nanometers, nm) through the shortest visible light wavelength (about 380 nm). It even measures some invisible (to you) light in the ultraviolet (below about 380 nm). This instrument, called a spectrophotometer, will be set up and operated by your laboratory instructor. You are responsible for giving your sunglasses to your lab instructor (labelled in some way for identification) and getting back a copy of the spectral

 "through put," the spectrum of your sunglasses. You are also responsible for getting a copy of the spectrum of a pair of ordinary plastic lab glasses. This spectrum may be run during the experiment or may be run prior to the lab, copied, and handed out during the lab period.

After you obtain a copy of the data for your sunglasses and the plastic lab glasses, label the axes of both of the plots: the X axis is wavelength measured in nm and the Y axis is transmission. Ask your lab instructor if you are not sure about the axes labels.

Report Sheet   Experiment 19

Ultraviolet Radiation and You

Date _____   Section number _____   Name _____

1. Data

     a) Staple or paper clip the spectrum of your sunglasses to the back of this report sheet.

     b) Staple or paper clip the spectrum of the plastic lab glasses to the back of this report sheet.

     c) Does your sunglasses' manufacturer make claims about your glasses' ability to transmit (or block) light? If so what?

2. Approximate light transmission

     a) What percentage of the visible light gets through the lab glasses? .... _____ %.

     b) What percentage of ultraviolet light get through the lab glasses? .... _____%.

     c) What percentage of the visible light gets through your sunglasses? .... _____ %.

     d) What percentage of ultraviolet light gets through your sunglasses? .... _____%.

     e) Which pair of glasses does a better job of blocking UV?

3. Colored sunglasses transmission characteristics (complete only if your sunglasses have obviously tinted lenses)

a) How are the colors *that you can see* through you sunglasses' lenses related to their spectrum? (Describe the shape of the spectrum's curve or line.)

b) How are the totally colorless lenses of the plastic lab glasses related to *their* spectrum?

Ultraviolet Radiation and You

Date _____   **Section number** _____   **Name** _____

1. Where *does* the ultraviolet light that hits the earth come from? Are there any other extraterrestrial sources?

2. What molecule or what cycle involving this molecule in the stratosphere removes a large amount of the incoming ultraviolet radiation?

3. What percentage of the ultraviolet light shining on your particular pair of sunglasses *is blocked by them*? How is this blocking achieved? (Hint: see introduction.) If you have a manufacture's claim about UV blocking ability, was it true?

4. What percentage of ultraviolet light shining on a pair of ordinary plastic lab glasses *is blocked by them*?

5. Did the cost of a pair of sunglasses seem to affect their ability to block ultraviolet light? Can you think of a reason that this would be so? What is the reason?

Think, Speculate, Reflect, and Ponder

6. If a result of the Chapman Cycle is that the energy of incoming ultraviolet radiation gets converted into infrared radiation in the upper stratosphere, what happens to the temperature of the upper stratosphere because of this process? (Hint: What is the relationship between infrared energy and heat?)

7. What recent stratospheric changes might suggest that the temperature of that atmospheric layer could change in the future?

8. There is lots of infrared radiation in the troposphere. Using Table 19.1 in this lab's introduction, give a reason why infrared radiation does not attack chlorofluorocarbons like stratospheric ultraviolet light does.

9. What will the visible light transmission spectrum of a red-tinted pair of sunglasses look like? Draw this spectrum and label the axes.

Experiment 20

The Apparent Molecular Weight of Air

Sample From Home

No samples from home are required for this experiment.

Objectives

The objective of this experiment is to allow you to measure the molecular weight of air by treating this complex mixture of a number of gases as if it were a pure gas. You will use an elegant and simple procedure that capitalizes on the sensitive ability of a three- or four-place balance to detect small changes in mass.

Background

The densities of the solids and liquids that you examined in Experiment 2: *Going Metric With the Rest of the World* are not strongly affected by changes in the physical conditions of the surroundings. The metal bar—for which you measured the density—in that experiment does not significantly vary with temperature unless you heat it to high temperatures and melt it!

The situation for gases is significantly different: The density of a gas depends strongly on its temperature and pressure. A mathematical equation that describes this relationship is

$$d = \frac{PM}{RT}$$

Where d = density of a gas; P = pressure; M = molecular weight; R is the gas constant; and T is Kelvin temperature.

Since the molecular weight, M, of a particular gas and the gas constant, R, are constants, the density of that gas is proportional to pressure and inversely proportional to temperature. This is a situation quite different from the effects of temperature and pressure on solids and liquids.

This result has profound effects on our weather. The positions of high and low pressure centers cause dramatic air movements because of the differences in air density. We all know that hot air rises; hot air balloons operate on this principle. This is also the reason for rapid mixing in the troposphere, the layer of the atmosphere closest to the earth. Air heated at the surface of the earth is mixed with other air masses as it rises to higher altitudes. This mixing, in effect, cleans the air by diluting (with clean air from other parts of the troposphere) the pollutants that were injected into the air at the earth's surface. During smog alerts there is often an absence of air movement due to an inversion layer. This occurs when warm air lies above cold air, thereby trapping the cold and increasingly polluted air next to the surface of the earth. Due to density effects (colder, denser air stays low and warmer, less dense air stays high) very little air movement and mixing occurs during a temperature inversion.

The effects of inversion layers upon tropospheric air quality became clear in the Los Angeles basin as early as the 1940s. Programs to clean up the air in the region were started then. Unfortunately population growth, more automobiles, and more industry have continued to outpace the remedial actions. A similar situation now exists in the Denver metropolitan area where an ugly "brown cloud" is highly visible on days when a temperature inversion is present. From the front steps of the National Center for Atmospheric Research, on a mesa above Boulder, Colorado, a layer of extremely dark polluted air can be seen often during the mid to late winter months as polluted urban air from Boulder and Denver mix together to obscure that beautiful mountain setting. The scene over Houston, Texas, is much the same.

The very slow rate of vertical mixing in the stratosphere (the layer of the atmosphere above the troposphere) is caused by a natural temperature inversion. As air rises in the atmosphere, it expands and cools. As a result, in the troposphere temperature tends to decrease with increasing altitude. When one reaches the stratosphere, however, this trend is reversed because of the presence of ozone. Ozone molecules absorb visible and ultraviolet radiation from the sun. This absorption of radiation heats the upper stratosphere so that the upper (stratospheric) air layers become hotter than the lower layers. The stratosphere, therefore, gets warmer as altitude increases, just the opposite of the troposphere. This also suggests that as the ozone layer is eroded by anthropogenic chemicals (made by humans) like chlorofluorocarbons, the temperature of the stratosphere will change—an hypothesis that has been born out in the past few years.

The cold air near the bottom of the stratosphere is relatively dense so it has little tendency to rise. The warmer air near the top of the stratosphere is less dense and has little tendency to sink. The result is that once pollutants enter the stratosphere, they tend to remain there for a very long time. The residence time (the average time present) of a pollutant in the stratosphere is known from tests of nuclear weapons in the atmosphere (before the nuclear test ban treaty of 1964; see Experiment 4: *Radioactivity*). It was found that the residence time of radioactive debris in the atmosphere ranged from about six months when injected low in the stratosphere (near 12 km) to about five years when injected high in the stratosphere (near 45 km).

The molecular weight of a compound is dependent on the atoms that make up that compound. This means that a mole (or a certain volume) of a particular compound that contains only one kind of molecule has a well-defined mass because only those kinds of molecules are present. However, for a mixture of compounds, the total mass of these molecules is made up of contributions from the masses of each of the individual molecules that are present. Just like the apparent (average) molecular weight of the elements reported on the periodic table is made up of contributions from the masses of the individual isotopes that are present, this experiment is entitled the Apparent Molecular Weight of Air because *air is not a pure compound containing only one kind of molecule.*

In fact air contains many different molecules; however, it is mostly made up of nitrogen, oxygen, argon, and carbon dioxide. All of these molecules together contribute to the mass of a particular sample of air. If we know the volume of this sample and perform some tricky yet relatively simple manipulations of the temperature of this sample, we can derive the apparent molecular weight of the sample. For a pure gas sample (containing only one kind of molecule, like methane or carbon dioxide), this experiment would yield the molecular weight of that pure compound:
CH_4 16.04 g/mole or CO_2 44.00 g/mole.

In this experiment you will make a number of observations of the mass and volume of a gas sample at two different temperatures. Be as careful as possible in measuring the mass and volume during the experiment. The sealing off of the gas sample in an Erlenmeyer flask at one point in the procedure requires the use of a tubing clamp to pinch the rubber tubing closed. Make sure that this clamp is tightly clamped at the proper place on the tubing. This is the most crucial step in the whole procedure. Ask for help with the balance if you have not been instructed in its use. You will perform this experiment in duplicate and average your results together.

An example calculation is included *after* the report sheet to help you perform the calculations for your experimental data correctly. This example data computes the molecular weight of gas from a Bunsen burner.

Procedure[*]

1. Assemble the apparatus as demonstrated by your lab instructor and as shown in Figure 20.1.

Use a small amount of glycerine to aid in the insertion of the small piece of glass tubing into the one-hole stopper (# 6). Insert a 3 inch piece of glass tubing ($^1/_8$ inch inside diameter) into a one-hole stopper. Make sure that the glass tubing does not extend below the bottom of the stopper. Put a 2 $^1/_2$ inch piece of rubber tubing on the end of the glass tubing that sticks out of the top of the rubber stopper. Make sure that the rubber tubing diameter is small enough that it makes a tight

[*] For additional information see "The Density and Apparent Molecular Weight of Air" by Harris, A.D. in *J. Chem. Ed.* 61(1), 74-75(1984).

seal to the glass. Place a tubing pinch clamp (or screw clamp) snugly *about the glass tubing* so that air can easily escape from the flask as you heat up the apparatus in Step 5. Do not clamp the rubber tubing at this time. Put the stopper *snugly* into the mouth of a 250 mL Erlenmeyer flask.

2. Weigh the entire apparatus (250 mL flask, stopper, tubing, and clamp) and record the weight to at least 0.001 g (or to four decimals if your balance is able). Clamp a large iron ring to a ring stand. Put a piece of wire gauze with a ceramic center on the iron ring. Place an 800 or 1000 mL beaker on the gauze and clamp it to the ring stand if a beaker clamp is available.

3. Using a large buret clamp, clamp the neck of the Erlenmeyer flask so that it is down inside the beaker as much as possible.

4. Fill the beaker (containing the clamped Erlenmeyer) with water almost to the beaker's lip, allowing only enough room (about ½ inch) for bubbling due to rapid boiling when you heat the water.

Figure 20.1. Apparatus during first ten minutes of boiling with rubber tubing unclamped.

5. Record the temperature of the water correctly to 1 ° C. Light a Bunsen burner with a match; adjust the flame to a clean, hot, blue flame (see Experiment 1: *The Ubiquitous Bunsen Burner*). Heat the water to a boil and then boil for 10 minutes.

6. After 10 minutes of boiling, record the temperature of the water to 1° WITHOUT ALLOWING THE TIP OF THE THERMOMETER TO TOUCH ANY OF THE GLASS (EITHER THE EDGE OF THE BEAKER OR THE ERLENMEYER).

7. With the water still boiling, use the pinch clamp to close the rubber tubing tightly. Make sure that the clamp is firmly closing off the rubber tubing's entire diameter. See Figure 20.2. This traps the air inside the volume of the flask.

8. Carefully raise the closed Erlenmeyer system out of the boiling water. Leave the beaker stationary on the wire gauze.

9. Place the Erlenmeyer flask, still tightly clamped, on a folded paper towel and dry it thoroughly with a second paper towel or Kimwipe. DO NOT REMOVE THE CLAMP.

10. After the Erlenmeyer cools to room temperature, weigh it to at least 0.001 g WHILE THE RUBBER TUBING IS STILL TIGHTLY CLAMPED.

After you get the weight of the closed, room temperature system, you can test to see if you were successful in your attempts to clamp off the system and trap the air inside. After you have recorded on the report sheet the mass of the clamped, dry Erlenmeyer, listen carefully as you unclamp the rubber tubing. If you were successful you will hear a satisfying hiss as air rushes into the flask. If you do not hear a hiss then you might have had a leaky experimental apparatus (clamped tubing or rubber stopper), and you will probably have to repeat this particular trial.

11. In order to determine the volume of the flask apparatus, remove the stopper from the mouth of the Erlenmeyer and fill the flask with water.

Figure 20.2. Apparatus just before removal from boiling water with rubber tubing clamped.

12. Push the stopper back into the flask, forcing water to fill the glass and rubber tubing. With the stopper still in place, dry off any excess water.

13. Pour *all* of the water from this system into a 500 mL graduated cylinder and record the volume of the apparatus (in liters) in the data table on the report sheet. You should be able to read the volume accurately to 5 mL.

14. Repeat the entire procedure from Step 1. Make sure that you start your second trial with a room temperature apparatus when you are doing the first weighing.

15. Get the atmospheric pressure in the laboratory from the lab instructor who should report this value to you in units of atmospheres (atm). The temperature readings that you record on the data sheet in degrees Celsius must be converted to Kelvin for the calculations. This is easily accomplished by adding 273.15 to all of your Celsius readings. Record these Kelvin temperatures in the data table in the appropriate place. Look at the sample calculation performed on the page <u>after</u> your report sheet for help with your calculations.

Report Sheet   Experiment 20

The Apparent Molecular Weight of Air

Date _____   Section number _____   Name _____

1. Data table

	Trial # 1	Trial # 2
Initial Mass of Apparatus	g	g
Final Mass of Apparatus	g	g
Change in Mass	g	g
Final Temperature of Water, T_2	°C K	°C K
Initial Temperature of Water, T_1	°C K	°C K
Change in Temperature	o	o
Volume of Apparatus (air)	L	L
Atmospheric Pressure Today	atm	atm

2. Calculations (all calculations involving temperatures use Kelvin)–See example next page.

   a) Calculate the mass of air in the system (at T_1)   Trial # 1   Trial # 2
   (*use Kelvin   Temperature* !)
   (multiply the change in mass times T_2
   then divide the result by (T_2 minus T_1)) = ... ____ g   ____ g

   b) Calculate the density of the air (at T_1)
   (divide the mass of air, 2(a) by
   the volume of air from the data table) = ... ____ g/L   ____ g/L.

   c) Atmospheric pressure in the laboratory
   (in atmospheres from data table) .... _____ atm.

   d) Temperature in laboratory (Kelvin) .... _____ K.

   e) Calculate the apparent molecular weight of air
   (multiply the density of air 2(b) by 0.08206;
   multiply this result by the lab temperature 2(d);
   divide the result by the air pressure in the lab 2(c) = ... ___ g/mole ___ g/mole.

e) Average of your two apparent molecular weights .... _____ g/mole.

Example Calculation

Bunsen Burner Gas

1. Example data table for the gas coming from your laboratory gas lines

	Trial # 1	Trial # 2
Initial Mass of Apparatus	148.551 g	g
Final Mass of Apparatus	148.516 g	g
Change in Mass	0.035 g	g
Final Temperature of Water, T_2	100.0 °C	°C
	373.15 K	K
Initial Temperature of Water, T_1	25 °C	°C
	298.15 K	K
Change in Temperature	75.0 °	°
Volume of Apparatus (air)	0.265 L	L
Atmospheric Pressure Today	1.010 atm	atm

2. Example calculations (all calculations involving temperatures use Kelvin)

a) Calculate the mass of air in the system (at T_1)   Example Trial

(*use Kelvin Temperature!*)
(multiply the change in mass times T_2
then divide the result by (T_2 minus T_1)) =

$$\frac{0.035 \times 373}{75}$$

0.174 g.

b) Calculate the density of the gas (at T_1)
(divide the mass of air, 2(a) by
the volume of gas from the data table) =

$$\frac{0.174}{0.265}$$

0.657 g/L.

c) Atmospheric pressure in the laboratory
(in atmospheres)

1.010 atm.

d) Temperature in laboratory (Kelvin)

298 K.

e) Calculate the apparent molecular weight of gas
(multiply the density of air 2(b) by the gas constant
0.08206; multiply this result by the temperature
in the lab 2(d) in Kelvin and divide the
result by the air pressure in the lab 2(c) =

$$\frac{0.657 \times 0.08206 \times 298}{1.010}$$

15.9 g/mole.

The Apparent Molecular Weight of Air

Date _____   **Section number** _____   **Name** _____

1. Would this experimental procedure work for the calculation of the molecular weight of a pure gas? How could you modify the procedure to accomplish this?

2. If you performed this procedure on the top of Mount Everest, would the apparent molecular weight of air that you determined be the same as if you performed this procedure at sea level? (Hint: Reread the background in this experiment and Experiment 5, O_2 *Content of Air.*)

Think, Speculate, Reflect, and Ponder

4. Why is the temperature of boiling water not 100 ° C at altitudes substantially higher than sea level? (An error in thermometer calibration or reading is not the answer.)

5. Why might the word "boiling" occur to astronauts just after their space suits rupture in space during a space walk?

Polymers

The World of Plastics and Synthetics

(The next three experiments illustrate the preparation of some commonly encountered giant molecules using the chemical process of polymerization. Some properties of these polymers will be examined. The nylon rope is really neat to see being made and the other two give you a souvenir that can be taken home.)

We should begin by answering the question, "Just what **is** a polymer?" A chemist might say that "A polymer is a molecule made out of monomers." But what, then, is a *monomer*, and how does it differ from any ordinary molecule?

Most molecules can be characterized as having a particular reactive site (*functional group* chemists would say) which, as the name implies, can react and combine with reactive sites in other molecules to form a single bigger molecule. In very general terms, we could diagram this as follows where the "hooks" represent potential chemical bonds due to these reactive sites.

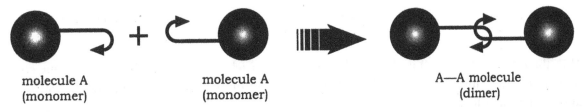

molecule A   molecule A   A—A molecule
(monomer)   (monomer)   (dimer)

The single molecules are *monomers* (mono = one), and the larger molecule would be called a *dimer* (di = two). But once we make the dimer, we cannot go any further since there are no more reactive sites available for joining together with still more molecules.

But what if one of the molecules had the potential to form not one but *two* chemical "hooks" per molecule? Then we could write:

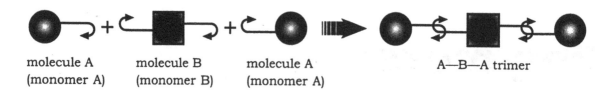

molecule A
(monomer A)
    molecule B
    (monomer B)
        molecule A
        (monomer A)
            A—B—A trimer

Now we have made a *trimer* (tri = three) by joining together a total of three monomer molecules, but here again the reaction must stop at this point since no more chemical hooks are left in the trimer.

You should now be able to guess the last chapter in this polymer story—how can molecules be designed so that they can continue to combine with each other almost indefinitely to produce a polymer? In order for this to happen, *all* of the monomer molecules must have the potential to form chemical bonds in both directions, i.e., have in effect *two* chemical hooks. Let's diagram a picture of what can happen in such a case:

monomer A   monomer A   A—A dimer

But the initial dimer itself still has two available hooks at each end, so:

A—A dimer   A—A dimer   A—A—A—A tetramer

Now the *tetramer* (tetra = 4) itself has still two hooks on each end which enable it to continue combining with more molecules in both directions with either more monomer molecules, or maybe another tetramer forming an *octamer*.

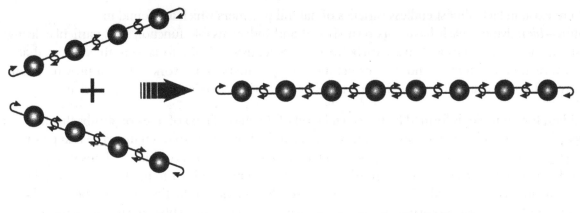

two A—A—A—A tetramers   A—A—A—A—A—A—A—A octamer

So where do we stop? Not until we use up all the reactants or "hooks" (monomer starting material), at which point we have a truly giant "mer" called a *polymer*. Monomers are thus simply molecules which may be identical or different, but all have the ability to form bonds in at least two directions. Polymers (poly = many) simply consist of usually thousands, or even millions, of monomer units all chemically joined together in one giant molecule.

Whereas most ordinary molecules have molecular weights (sizes) not greater than a few hundred, polymers typically have molecular weights from tens of thousands to many millions. Some of these large polymers are even big enough to see with special instruments like the electron microscope, although the individual atoms themselves still cannot be discerned. Most polymerizations also require small amounts of chemical reaction initiators (catalysts) to cause the monomer molecules to react at a practical rate; other chemicals are added to control the length of the polymer chain.

But do not be too quick to give human beings credit for inventing polymers, for they were present long before we made the earth scene. For example, the fibrous part of plants is principally cellulose, a polymer of glucose that gives bulk to our diet but is indigestible in humans and thus provides us with no food value.

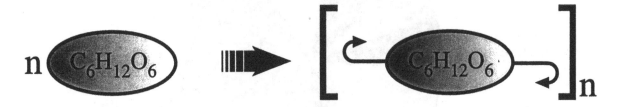

"n" molecules of glucose   cellulose polymer
                                    ("n" averages about 10,000 glucose units)

Starch too is a polymer—and also a polymer of glucose. We can, however, digest starch and obtain food value from it. If that sounds impossible or intriguing, perhaps you will decide to study organic chemistry in three dimensions to find out the answer.

There exist, in fact, almost endless varieties of natural polymers which are found in nature—*bio*polymers each having its own special and indispensable function in a particular living system, such as proteins and many carbohydrates. Scientists' initial efforts were thus directed at discovering, understanding and duplicating these biopolymers. This work led to improving upon them and synthesizing new ones designed to suit the various needs and wants of society.

Rubber, for example, is formed by the coagulation of the juice (latex) of trees or shrubs that are successfully grown in the Far East, South America, and Africa. As is so often the case, real progress results only in crisis when a necessary determination is added to resourcefulness. Thus when World War II drastically and dangerously cut into the supply of rubber to the United States, this determination, coupled with necessary financing, resulted in the production of a number of rubberlike substances, some of which were equal or superior to natural rubber for tires and many other special uses. The development of these various synthetic polymers indicated that it is not necessary to duplicate or even approximate the chemical structure of natural rubber to produce an elastic plastic. And serendipity, through the eyes of *clever* and *curious* scientists, also has led to the discovery of new polymers like Teflon.

Almost all articles commonly referred to as "synthetic" or "plastic" represent man-made polymers. Since most of these polymers come ultimately from petroleum, we can better appreciate the need to conserve our remaining petroleum resources for these purposes rather than burning these resources up as a fuel. Alternate energy sources *can* be developed, but once the world's petroleum resources are gone, so will the raw materials to make much of what we wear, build with, ride in, drink from, sit on, cook with, and otherwise depend upon for our accustomed "good" life.

Many persons feel that plastics, their virtues notwithstanding, have become too much of a good thing in our society. The necessary stability of these polymers *during* use makes it difficult to get rid of them *after* use. Plastics make up over 7% of all landfill waste by *weight*, a number that is

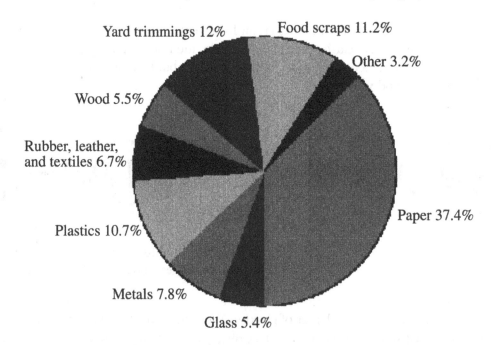

Total municipal solid waste generation (232 million tons) before recycling (2000).

steadily increasing. But these plastics comprise a huge 30% of all waste by *volume,* a figure often avoided by manufacturers and users of plastics. It is estimated to take upwards of 500 years for plastics to decompose in the ground, depending upon the kind of plastic, soil moisture and soil acidity.

How appetizing something is towards hungry microorganisms is a measure of what is referred to as its *biodegradability.* Microorganisms have a strong distaste for most plastics, never having "learned", through evolution, to digest them. Consequently, plastics are generally regarded as *nonbiodegradable*—certainly when compared to ordinary paper which biodegrades in just 2 to 4 weeks.

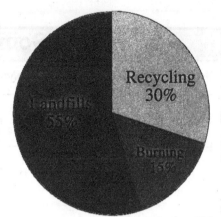

Disposition of municipal solid waste in the United States (2000).

Instead of consigning plastics to landfills, they can be burned; however, toxic vapors are often produced. This leads to a third option (after burying or burning): recycling. But alas, there are problems here too. The many different and incompatible plastics produced must be kept separate for recycling purposes. Worldwide, over 60,000 plastics are manufactured, although just six of these constitute 70% of the United States production.

A voluntary plastics numbering system has been instituted to facilitate the sorting and recycling process, but as many consumers are aware, the recycling of plastics often remains difficult. In most locales plastics, marked #1 PET (polyethylene terephthalate) or #2 HDPE (high density polyethylene) inside a triangle of chasing arrows, are recyclable. Much PET ends up in polyester carpets and the HDPE in plastic pipe, lumber and the manufacture of trash cans.

Plastics made up an estimated 390,000 tons of municipal solid waste generation in 1960. The quantity has increased relatively steadily to 24.7 million tons in 2000. As a percentage of MSW generation, plastics were less than 1 percent in 1960, increasing to 10.7 percent in 2000. While overall recovery of plastics for recycling is relatively small (1.3 million tons or 5.4 percent), recovery of some plastic containers is more significant. For the year 2000, PET soft drink bottles were recovered at a rate of 34.9 percent and high-density polyethylene milk and water bottles at 30.4 percent.

As with everything, there is always a risk/benefit analysis to its production and use. But the wise utilization of our resources can indeed provide us with many benefits to our life on this planet. Certainly, the development of the whole field of polymer chemistry has given us a wide variety of elastic and nonelastic materials which the consumer today takes for granted and would not want to give up. Some common synthetic polymers appear on the next page.

Some Vinyl Polymers

Monomer	Monomer formula	Polymer formula & name	Trade name	Uses
Ethylene	(H₂C=CH₂ structure)	(polymer chain) polyethylene	Polythene	Bags, film, toys & molded objects. Electrical insulation.
Propylene	(H₂C=CHCH₃ structure)	(polymer chain) polypropylene	Nalgene Herculon	Bottles, films, outdoor & indoor carpets; car seat covers
Vinyl chloride	(H₂C=CHCl structure)	(polymer chain) polyvinyl chloride	PVC Dynel Koroseal	Floor tiles, phonograph records, raincoats, garden hose, vinyl car tops
Tetrafluoroethylene	(F₂C=CF₂ structure)	(polymer chain) polytetrafluoroethylene	Teflon	Pan coatings, electrical insulation, gasket bearings, pipe tape.
Vinylidene chloride	(H₂C=CCl₂ structure)	(polymer chain) polyvinylidene chloride	Saran	Food wrap
Acrylonitrile	(H₂C=CHCN structure)	(polymer chain) polyacrylonitrile	Orlon	Fabrics & rugs

A Condensation Polymer

tetraphthalic acid monomer + ethylene glycol monomer → polyester copolymer (Dacron, Mylar) [Dacron thread used in fabrics & clothing; Mylar film used for magnetic recording tape and frozen food packaging.] + water

268

Experiment 21

A Polyamide: Nylon

Sample From Home

No samples from home are needed for this experiment.

Objectives

You will mix two chemicals together (monomers) and produce a nylon polymer which can be drawn out into a pliable "rope." You can then measure the overall length of the nylon "rope" you have made from one fourth cup of liquid.

Background

One common way to join molecular units together to produce giant molecules is called *condensation polymerization*, where a small molecule like H_2O or HCl is "split out" and eliminated during the chemical reaction. A classic example of this is one for which the DuPont Company spent hundreds of thousands of dollars in the early pioneering days at the dawn of polymer research—money spent in the hope that something practical and profitable for the Company might be forthcoming. After initial discovery, millions of dollars had to go into the development of suitable manufacturing processes before the new polymer could be mass produced for consumer use. In the case of nylon, DuPont's gamble paid off.

In practice, nylon fiber is actually spun from a melt and pulled to give it desirable tensile strength. Although there are different nylon type polymers depending upon the specific monomers used, the reaction in this experiment is illustrative (see top of next page).

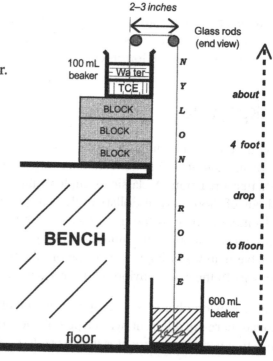

The reaction equation labels:

"n" molecules of sebacoyl chloride + "n" molecules of 1,6-hexanediamine → "n" units of nylon polymer chain + "n" molecules of hydrogen chloride

Nylon is thus a *copolymer* made by the chemical combination of *two different* monomers. The boxed groups of atoms at each end of the two monomer molecules represent the two reactive sites—our chemical hooks (see *Polymers: World of Plastics*). Molecular weights of the polymer molecules typically average around 15,000 where n = 52 units.

Each of the reactants is dissolved separately in two liquids that do not dissolve in each other (water and tetrachloroethylene); chemists call such liquids *immiscible*. When these two solutions are brought together, the monomers can react with each other only where the two liquids are in contact (called the *interface*). It is from this polymer film that forms at the *interface* between these liquids that the nylon rope is drawn. As soon as old polymer film is removed, new film immediately forms behind it, and so the reaction results in the continuous creation of more and more polymer until all the reactant monomers are consumed.

Procedure

CAUTION: DO NOT BREATHE VAPORS NOR LET ANY OF THE CHEMICALS IN THIS EXPERIMENT CONTACT YOUR SKIN; THE DIAMINE IS ESPECIALLY TOXIC.

If an accident occurs, immediately wash the contaminated area with soap and water and notify your lab instructor. Good ventilation is necessary in the laboratory for this experiment; make sure that the ventilation hoods are working well before you start this experiment.

Using your engineering aptitude, position necessary beakers and glass rods as shown in the diagram and pictures. The rods can best be mounted by inserting one end of each into a cork or stopper and gripping with a burette clamp attached to separate ring stands. They should be angled to form a very shallow "V" over which the nylon "rope" will be pulled (see Figures 21.1 and 21.2).

Figure 21.1. Nylon "rope" set-up.

Fill a 600 mL collection beaker about ⅔ full with water and place the beaker onto the floor. To get the desired height of fall, you will probably also have to raise the height of the beaker in which the polymer is made about a foot above the bench top with blocks or other suitable props.

USE DISPOSABLE PLASTIC GLOVES WHILE DOING PROCEDURE STEP 1.

1. The nylon rope monomer solutions have already been made for you. Pour 50 mL of the *sebacoyl chloride in tetrachloroethylene* solution into a 100 mL beaker. Very carefully and slowly layer 25 mL of the *1,6 hexanediamine in water* solution on top of the tetrachloroethylene solution so that a MINIMUM OF MIXING of the two layers occurs. Do not stir or agitate the beaker contents. It may help to pour the water solution down a funnel with the tip held against side of beaker just above surface of the tetrachloroethylene layer to avoid mixing.

(a) Note observations on the report sheet.

Figure 21.2. Correct engineering of set-up important for continuous "rope" production.

If you wish, you can dye your polymer some neat colors by adding safranin 0 (glow-red) or orcein (purple) to the 100 mL reaction beaker. Only a BB sized pinch dropped in the top liquid layer is needed. Do NOT use too much dye or the polymer "rope" will likely break before it can be pulled from the reaction beaker.

 Note: if you have not already had your engineering construction approved, HAVE THE LAB INSTRUCTOR CHECK YOUR SET-UP AT THIS POINT BEFORE PROCEEDING.

Now we are ready for the actual "rope" production! With gloves still on your hands, use tweezers to reach down through the top water layer and grasp the center of the *interfacial polymer film* and slowly pull the mass up through the water layer. Bring the nylon "rope" up from the center of the beaker, over the shallow "V" in both the glass rods as shown in the diagram and then, still holding with the tweezers, guide it down into the large beaker on the floor. With a little experimentation and teamwork, the height of the nylon rope falling over the glass rod can be adjusted so that the weight of this "rope" will be sufficient to keep pulling more rope automatically from the solution. Time to just stand back and watch. To avoid tangling, some students try winding the rope up on the outside of a large beaker as it forms.

Avoid touching the nylon rope as much as possible to prevent the reagents from unnecessarily contacting your gloves. If you do get the chemicals on your hands, wash up with soap and water and contact the lab instructor.

When the polymerization is finished, wash the polymer rope in your 600 mL collection beaker in cold running water for a couple of minutes. Drain off all of the water, add a mixture of 25 mL water plus 25 mL of acetone and agitate gently to wash the polymer thoroughly with this solvent. Pour off the acetone/water solution into an appropriate waste container, refill the beaker with water and gently lift out your nylon "rope" and stretch it out on the bench top. Pour the tetrachloroethylene layer remaining in your 100 mL reaction beaker into the lab container labeled "Waste Tetrachloroethylene."

Figure 21.3. Nylon "rope" in production.

 After washing with water and while the rope is still in the 600 mL beaker, try sticking one end of the rope onto the table top. Then walk along with the beaker and simultaneously pull the rope out and onto the bench top with your hand. It takes a lot of care, patience, technique and sometimes luck to be able to untangle your rope. If your rope just won't untangle in a reasonable length of time (20 minutes), cut it into pieces sufficient so you can at least estimate its length. One hundred feet is an excellent "score."

2. Perform the following tests on the nylon rope:

(a) Describe the color, appearance, texture, shape and tensile (breaking) strength of your damp nylon rope.

(b) Spread out your polymer on the desk top in such a way that you can measure its total length with a meter stick. Record your results on the report sheet.

(c) Test the inertness (non-reactivity) of your polymer towards the solvent acetone by placing a pea-sized polymer wad into about 10 mL of acetone in a 50 mL beaker. CAUTION: Acetone is flammable. Extinguish any flames near work area. If the nylon becomes sticky, this indicates that it is being dissolved by this solvent.

Figure 21.4. Finished nylon "rope" laid out on the bench top.

Note your observations on the report sheet and dispose of the liquid and the polymer wad in a suitable waste receptacle. (The nylon will, however, become very brittle when dry since no plasticizer chemical was added to keep the rope flexible.)

A Polyamide: Nylon

Date _____ **Section number** _____ **Name** _____

1. Mixing of the Monomers

    (a) Observations when sebacoyl chloride and 1,6-hexanediamine solutions come into contact.

    _____

2. Nature of the "Rope"

    (a) Physical properties of nylon rope:

    Color_____.

    Texture_____.

    Shape_____.

    Tensile strength_____.

    (b) Total length of rope from 40 mL of solution. (If your lab instructor agrees, stretch it out around the lab!)

    _____meters.

    _____ feet.

    (c) Solvent resistance of polymer towards acetone_____.

3. Comments and conclusions on experiment

A Polyamide: Nylon

Date _____   **Section number** _____   **Name** _____

1. What do you think might be the purpose of washing the newly formed polymer with an acetone/water solution?

2. List three principle consumer uses for nylon.

(a)

(b)

(c)

Think, Speculate, Reflect and Ponder

3. Household utensils made of plastic are now a commonplace item. Mention some household items that have not been successfully fabricated from plastic. Do you expect that any of these articles you have mentioned can or will be made out of plastic in the future? Why or why not?

4. Regarding polymers, why doesn't it make sense to burn up all the petroleum reserves for energy before switching to alternate energy sources?

Experiment 22

Polystyrene: Casting Resin

Sample From Home

If you choose, you may bring any small object (picture, ring, agate, coin, bug) that you would like to embed in plastic.

Objectives

We will continue our investigation of polymers by making polystyrene, an addition polymer. You have the option of embedding an object of your choice in this piece of solid, clear plastic.

Background

Along with polyethylene (Tupperware®, plastic tarpaulins, etc.), polystyrene is one of the most widespread of all man-made plastics, due in large part to the fact that both of these polymers are relatively cheap to produce. Polystyrene can be molded into either a hard solid (this experiment), or a light, foamed shape. Such molded polystyrene objects might also be termed "cheap" in another sense due to the ease with which they crack and break over time. This cheapness however, makes polystyrene economical even as a paper substitute for some applications where both food (e.g., meats at the supermarket) and breakable objects are packed using foamed polystyrene.

In addition to its significant bulk in landfill waste and relative non-biodegradability, until the mid 1990's styrofoam manufacture put tremendous amounts of methylene chloride into the atmosphere. (Methylene chloride was used as a foaming agent but represents a potential long term danger to the ozone layer — a chemical shield high in the stratosphere that protects life on the surface of the earth from harmful ultraviolet radiation.) But some competitors to styrofoam might surprise you. Where else but in the United States could one find packages being shipped using food instead of foam or even crumpled newspaper as a filler: popped popcorn and water soluble, foamed starch that can be washed down the drain.

Unlike nylon, polystyrene (as well as polyurethane in the next experiment) is produced simply by the joining together of monomer molecules without splitting out any small molecules in the process. This process is called *addition polymerization:*

"n" molecules of styrene
(monomer)

section of polystyrene polymer chain
("n" units)

Each hexagon in these structures represents a ring of six carbon atoms. Molecular weights of the polymer molecules may range up to 500,000.

The chemical 2-butanone peroxide serves as a catalyst to initiate the polymerization so that it will proceed at an acceptable rate. Because timing is critical, you must pay close attention to the mixing and setting times for this reaction as mentioned in the procedure.

Procedure

Note: Good ventilation is necessary for this experiment; you will need to mix and react the styrene *monomer* in a HOOD. Because of cleanup difficulties, we will avoid getting these chemicals on the lab glassware.

Determine a 20 mL volume in three ounce waxed paper cup by pouring 20 mL of water into one cup and marking the water level on the outside with a pen or pencil. (This process is called *calibrating* your cup.) Use this mark to calibrate two empty *dry* cups at the 20 mL volume level and then either discard your cup plus water or let someone else use it for their calibration.

CAUTION: USE DISPOSABLE GLOVES WHEN HANDLING STYRENE MONOMER IN PARTS 1 AND 2. PERFORM BOTH PARTS 1 AND 2 IN A HOOD.

1. Fill just one of your calibrated cups to the 20 mL mark with styrene monomer and add the number of drops of catalyst specified by your laboratory instructor. Mix thoroughly by swirling the cup contents for two minutes.

TAKE CARE THAT ABSOLUTELY NO TRACE OF CROSS CONTAMINATION BETWEEN THE STYRENE AND PEROXIDE STOCK CONTAINERS OCCURS WHEN YOU GET YOUR SAMPLE.

If you choose to stir, do so slowly to minimize incorporating air bubbles into your liquid. If you wish to have *color* in this one half of your plastic block, use a tiny spatula or wood splint to add a "three pin heads" size pinch of Sudan III or IV red dye to

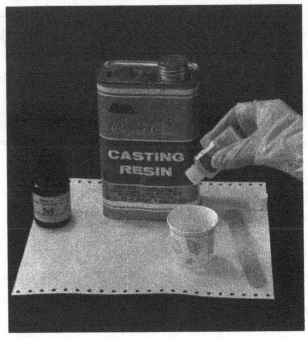

Figure 22.1. Adding catalyst to styrene monomer.

your monomer solution at this point. Too much dye will retard the curing of your plastic block. (The names of these dyes should be color explanatory if not chemically meaningful.) When you brew up the second batch of styrene monomer in Part 2, you can decide whether to add color to it or just leave it *clear* (usually a better choice).

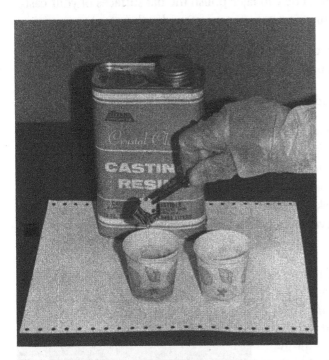

Figure 22.2. Embedding coin in resin.

A typical gel time is 10 minutes @ 22 °C.

(a) Record the room (catalyst) temperature on the report sheet.

(b) Note the time of catalyst addition and record the total minutes elapsed before gelling begins.

(c) Note the viscosity ("flowability"), color and (cautiously) odor of the monomer solution at this point.

Select some object of your choice to embed if you wish, such as a picture, ring, agate, coin, bug, etc. Leaves and flowers can be preserved, but they must be specially treated first to remove water and air from the tissue in order to obtain a perfectly clear block.

2. About five minutes after you mix the first monomer batch, mix a *second* batch of styrene monomer plus catalyst in your *second* calibrated paper cup (color may be added as before, if desired). Drop your object to be embedded into this *second* batch. When the *first* batch has gelled enough to support the object you plan to embed, remove the object from the *second* batch with tweezers and place it on top of the *first* batch in the position you want, top down (Figure 22.2). Then pour the *second* batch of resin on top of the first (Figure 22.3). Remove any air bubbles with the aid of a toothpick or wire. The *second* batch of resin should become

Figure 22.3. Pouring second batch of resin.

sufficiently hard in 30–45 minutes after the initial mixing to remove the whole plastic block from the cup.

NOTE: The resin cup should be warm to the touch during hardening. If it gets too hot to touch comfortably, place the lower $\frac{1}{2}$ of the cup into some cold water for a few minutes; otherwise, excessive heat may cause cracks in your casting. Or, if the resin is not hardening quickly enough, try putting the lower half of the cup into some hot water.

Describe your observations on the report sheet. (You can later polish the flat surfaces of your casting at home by rubbing it over a piece of very fine emery paper moistened with water.)

3. Although most of the curing/hardening process is not complete for several hours (like concrete, it never actually completely stops), we can still proceed with the following tests forthwith:

(a) Describe the hardness, texture, color and general appearance of your plastic casting.

(b) Test the inertness of your polystyrene plastic casting towards the solvent acetone (nail polish remover) by letting a few drops of acetone fall onto the surface of your casting. (CAUTION: *Acetone is flammable. Make sure that no flames are near the work area.*) Rub the plastic surface with your finger where the acetone has been dropped. If the surface becomes sticky, this indicates that it is being attacked (dissolved) by this solvent. Note your observations on the report sheet.

(c) Weigh your plastic resin casting and then determine its volume using the methods described in Experiment 2: *Going Metric with the Rest of the World.* You will have to obtain a graduated cylinder just large enough for your casting to slip into (probably a 1000 mL size). Use these data to calculate the average density of your *object + resin* casting.

Figure 22.4. Hardened resin casting.

Polystyrene: Casting Resin

Date _____   **Section number** _____   **Name** _____

1. Monomer/catalyst mixture used _____

    (a) Room temperature   _____°C.

    (b) Time required for gel   _____ min.

    (c) Physical properties of styrene liquid

        Viscosity_____

        Color _____

        Odor _____

2. General observations during preparation of casting

_____

3. Polystyrene casting

    (a) Physical properties

        Hardness _____

        Texture _____

        Color _____

    (b) Solvent resistance of polystyrene towards acetone _____.

    (c) Density of casting

        Weight (to nearest 0.1 g) .... _____g.

        Volume (to nearest 1 mL) .... _____mL (cm^3).

        Calculated density
        (divide the mass by the volume) = ... _____ g/cm^3.

4. Comments and conclusions on the experiment

Polystyrene: Casting Resin

Date _____   **Section number** _____   **Name** _____

1. You were cautioned not to permit the slightest contamination between the styrene and peroxide stock containers. What might happen if these precautions were ignored?

2. Would your plastic block float or sink if dropped into the Great Salt Lake in Utah? (Density = 1.1 g/cm^3). Show calculations for your sample to back up your answer.

3. Why does the price of oil affect the price of plastics?

4. In general terms, how do you think this procedure would have to be changed in order to make a *foamed* instead of *solid* cast block of polystyrene? (Hint: What is the difference between leavened and unleavened bread? Also, see this experiment's background discussion.)

Think, Speculate, Reflect and Ponder

5. Why were only very small amounts of 2-butanone peroxide required in this experiment? Hint: Look up the term *catalyst* (a term not to be confused with a rancher's inventory).

6. Would the procedure's methods that describe ways to change the temperature of the polymerization reaction vessel (the paper cup) also change the rate of the ongoing reaction? Explain.

Experiment 23

Polyurethane: Rigid Foams

Sample From Home

No samples from home are needed for this experiment.

Objectives

In completing our study of polymerization, we will investigate another type of addition polymerization reaction. However, instead of obtaining a solid block of plastic as with polystyrene, a very light, rigid foam will be produced.

Background

Although more expensive than polystyrene plastics, polyurethanes have found extensive application in home and industry. Not only are the foams relatively inert, but they also have good insulating properties which makes them attractive for use in refrigerator and freezer walls. And because of the extreme lightness and rigidity which can be given to the foam, lamination of this foam into hollow structural components can produce very strong, rigid, but light panels especially valuable to the aircraft industry.

Polyurethane formulations can furthermore be altered to give a nonrigid elastic foam, such as that commonly found in pillows and cushions. Spandex® elastic fibers are a urethane-type polymer. Other modifications can yield very durable imitation leather products (Corfam®) which have the ability to "breathe" air and water vapor through its pores like real leather. This is especially important in shoes. These are only a few of the many uses for this polymer, but they should be enough to indicate the importance and versatility of this synthetic "plastic."

The two *viscous* ("honey-like") solutions A and B used in this experiment contain the two monomer molecules (the polymerization thus produces a co polymer) and also the catalyst necessary to make the reaction proceed at a practical rate. Upon mixing, the hooking together of the monomers (the *polymerization*) begins. On a molecular level, this reaction is an addition polymerization like polystyrene and looks approximately like the following horrible mess when expressed in chemist's notation:

toluene diisocyanate

poly (propylene oxide) glycol
[this is a polymer itself]

polyurethane

The groups of atoms in shaded boxes represent the reactive sites (our chemical hooks) which serve to join the whole polymer together; each hexagon represents a ring of six carbon atoms.

But if this were all that happened, you would end up with just a solid, intractable glob. With no foaming agent, it would be like trying to bake a cake without any baking powder—no rising and foaming of the product would result. So just as cakes and breads rely on a chemical reaction to produce carbon dioxide gas for foaming, so does our polyurethane reaction. Some of these N=C=O groups that you see on the toluene diisocyanate molecules react with the water present to produce the CO_2 (carbon dioxide) necessary for frothing.

Although timing can be important in all chemical reactions, it is *critical* when carrying out polymerizations and especially so with polyurethane foam preparations. Different commercial formulations may vary somewhat in their specifications, but in general, mixing of the two solutions A and B must be completed within 30–60 seconds. After this time, foaming accelerates rapidly as the mixture expands to about *twenty times* its original volume. Foaming requires 3–5 minutes and most further chemical reactions are complete in 20–30 minutes. The full strength of the foam is not reached until 24 hours of cure time has passed. Just like the saying "Time waits for no one," once mixed, neither do chemicals!

286

Procedure

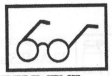

 CAUTION: THE ISOCYANATE MONOMER USED IN THIS EXPERIMENT IS VERY TOXIC—BOTH BY BREATHING AND SKIN CONTACT. Good ventilation is necessary in the lab for this experiment. And because of clean-up difficulties, we will avoid getting the monomer liquids or the polymer itself on the lab glassware.

USE THE DISPOSABLE PLASTIC GLOVES. ALL POLYMERIZATION WORK DURING STEPS 1 AND 2 MUST BE DONE IN A HOOD.

Determine a 25 mL volume in two 3 ounce waxed paper cups by pouring 25 mL of water into each of them and marking the water level on the outside of each cup with a pen or pencil. (This process is called *calibrating* your cup). Discard the water and blot out residual water with a dry towel.

1. While in the HOOD and using the plastic gloves provided, fill *one* of your two calibrated cups to the 25 mL mark with monomer A solution, and the *other* cup with monomer B. TAKE CARE THAT ABSOLUTELY NO TRACE OF CONTAMINATION BETWEEN A AND B CONTAINERS OCCURS WHEN YOU TAKE YOUR SAMPLES. Note the visible characteristics of the monomer solutions (viscosity, color), but do not smell the solutions!

2. Pour the less viscous (less thick) monomer into the other, scraping out the last bit of liquid with a wooden tongue depressor. Mix well and without delay (30–45 seconds), drop the cup containing the monomer mix into a one quart-size plastic bag or large disposable plastic glove with pinholes made in the tips of fingers to let the air escape. Note your observations.

3. After your polymer foam has set, cooked, and cooled (30–45 minutes), perform the following tests on your product:

(a) Describe the porosity (gas hole content), rigidity (compression resistance), texture, color, and general appearance of your polyurethane foam.

(b) Test the inertness of your polymer foam towards the solvent acetone (nail polish remover).

Figure 23.1. Mixing together polyurethane chemicals.

Since acetone is a flammable organic liquid, you must first MAKE SURE THAT NO FLAMES are near your work area before proceeding. Perform the test by placing a pea-sized piece of foam into a crucible and adding about 3 mL of acetone. If the foam collapses or becomes sticky, this indicates that it is being attacked by this solvent. Record your observations. Pour any remaining liquid directly into the sink drain and discard residual polymer into a suitable receptacle.

(c) Determine the density of a golf ball size piece of foam by first measuring its weight and then measuring its volume using the methods previously described for density determinations in Experiment 2: *Going Metric with the Rest of the World.* You will have to obtain a graduated cylinder just large enough for your foam to slip into. Use your ingenuity to submerge the foam and get an acceptably accurate measurement of the water displacement.

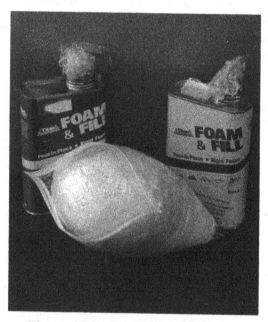

Figure 23.2. Finished foam block.

Polyurethane Foam

Date _____ **Section number** _____ **Name** _____

1. Appearance of

      Monomer Solution A _____ .

      Monomer Solution B _____ .

2. Observations upon mixing _____ .

3. Polyurethane foam

    (a) Physical properties

        Porosity _____ .

        Rigidity _____ .

        Texture _____ .

        Color _____ .

    (b) Solvent resistance of foam towards acetone _____ .

    (c) Density of polyurethane foam

        Weight of foam piece (to nearest 0.01 g) .... _____g.

        Volume of foam
        (to nearest 1 mL, or 0.1 mL if possible) .... _____mL (cm^3).

        Calculated density
        (divide mass by volume) = ... _____g/cm^3.

4. Comments and conclusions on experiment

Polyurethane Foam

Date _____   **Section number** _____   **Name** _____

1. Considering only the buoyancy effect, would you rather have a 1000 cm^3 block of polyurethane foam or a 1000 cm^3 block of cork (density = 0.2 g/cm^3) to hold onto in order to keep from drowning? Support your choice by showing suitable calculations.

2. Where do we obtain the raw materials to make most plastics?

3. How was your foaming produced? What gas was used, and where did the atoms making up the gas come from? (HINT: See the background discussion to this experiment.)

4. Consider the polymerization process itself:

(a) Directions for carrying out polymerizations usually refer to a *minimum* temperature for use. Why is that?

(b) What disadvantage might there be if the temperature were too hot?

Think, Speculate, Reflect and Ponder

5. Considering all of the polymers in this and the last two experiments, describe two of the difficulties that would be encountered in designing a plastics recycling program that requires the separation of each kind of plastic before the material could be recycled.

Experiment 24

Hangover Havens:
Salicylic Acid Derivatives
(Aspirin and Oil of Wintergreen)

Sample From Home

No samples from home are needed for this experiment.

Objectives

This experiment will demonstrate a simple synthesis of two well-known medicinal compounds, aspirin and oil of wintergreen, employing the techniques of crystallization, suction filtration and oven drying. The quantitative yield of aspirin product will be determined and the smell from oil of wintergreen enjoyed by all.

Background

LSD ("acid") is a dangerous drug enabling one to take a temporary but not very safe trip from worldly realities. Its hallucinogenic effects were accidently discovered in 1943 and can be produced by an oral dose as small as 50 micrograms (0.00005 g). Salicylic acid, on the other hand, was first prepared back in 1838, and by the late 1800s doses of 0.5 g or more were used not as an escape from reality, but as a relatively safe temporary escape from all-too-worldly headaches. In addition to its pain reducing (*analgesic*) effects, salicylic acid combats fever (*antipyretic*), relieves arthritic pains (*anti-inflammatory*), prevents heart attacks by thinning the blood and has a low toxicity with no LSD type flashbacks. It thus assumes considerable importance indeed as a drug to relieve the symptoms of several of our most common ailments.

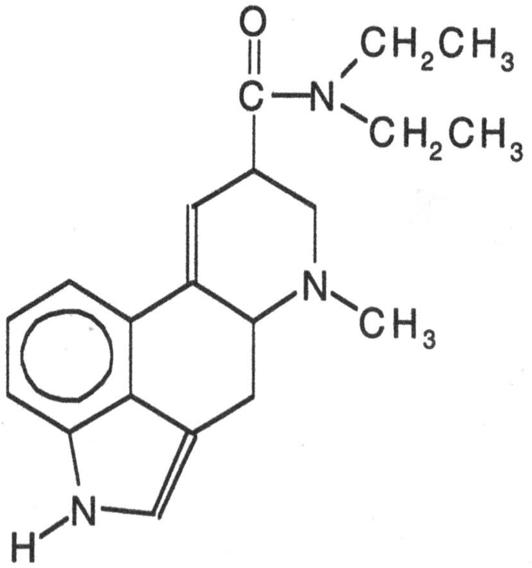

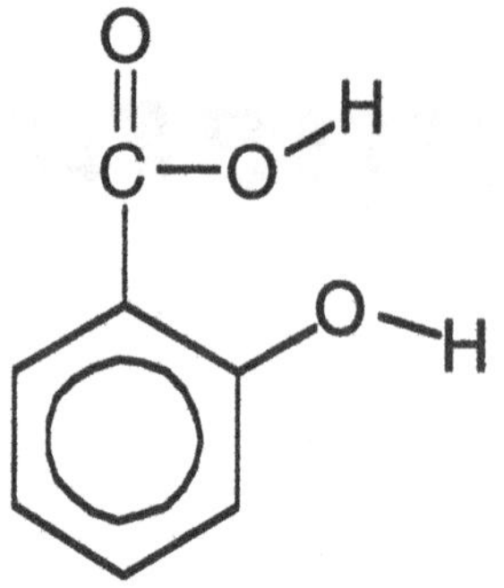

lysergic acid diethylamide (LSD)
(used by acid heads)

salicylic acid
(used for headaches)

(The lines in the above formulæ refer to chemical bonds, and each geometrical corner stands for a carbon atom unless otherwise indicated)

Its beneficial effects notwithstanding, salicylic acid also produces a number of unpleasant side effects in many individuals. For this reason, organic chemists sought to slightly modify its structure in a way to keep its beneficial properties while eliminating, or at least reducing, the undesired effects. This chemical modification is exactly what is practiced by chemists today—whether developing an optimum drug with a minimum toxicity or an optimum nerve gas of maximum toxicity. The molecular structure is altered by a combination of design and trial-and-error until a chemical having the desired properties is produced.

Aspirin

The particular chemical modification most widely utilized in the case of salicylic acid is called *acetylation*. This is one of the two chemical changes you will carry out in this experiment.

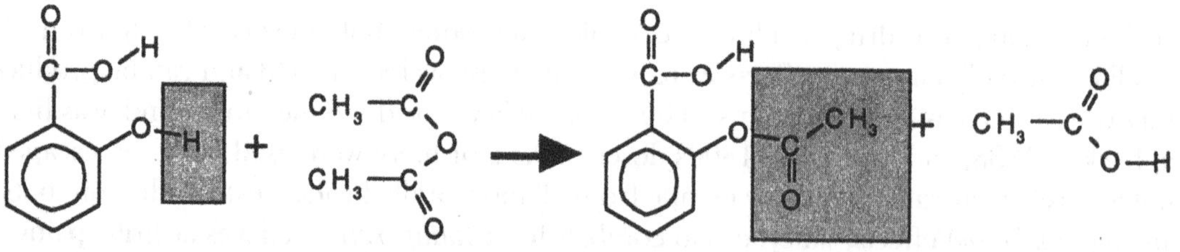

salicylic acid

acetic anhydride
(acetylating agent)

acetyl salicylic acid

acetic acid

As the shaded boxes in the equation structures indicate, this modification results in a hydrogen atom of the *hydroxyl* (OH) group attached to the ring on salicylic acid being replaced by an *acetyl* group $CH_3-C=O$. This chemically modified salicylic acid, called *acetylsalicylic acid* or more commonly just aspirin or ASA, was first put onto the market in 1899. The annual production of aspirin in the United States peaked in 1980 at 34 million pounds but has dropped since then, a decline undoubtedly due to health safety concerns and the availability of aspirin substitutes like acetaminophen (Tylenol®), ibuprofin (Advil®) and naproxen sodium (Aleve®).

Standard five grain tablets contain 0.324 g of aspirin, but in practice, less than half of the annual production of aspirin ends up in "pure" aspirin tablets. The rest is formulated along with other ingredients like buffers, special coatings and caffeine into a wide variety of products heavily advertised to be "better" than aspirin alone.

Brand Name ®	Ingredients
Anacin	aspirin, caffeine
Ascriptin	aspirin, buffer
BC fast pain relief	aspirin, salicylamide, caffeine
Bufferin	aspirin, buffer
Ecotrin	aspirin
Empirin	aspirin
Excedrin migraine	aspirin, acetaminophen, caffeine

Table 24.1. Some over-the-counter pain medications containing aspirin.

In spite of the hard sell come-ons that extol the virtues of a particular brand of aspirin or aspirin containing product, the best medical evidence still indicates that:

> Aspirin is aspirin—all brands must meet federal purity standards, so buy the cheapest brand of U.S.P. aspirin on the market. The Bayer Company does not, naturally, agree with this.

> None of the aspirin formulations is more effective than an equal amount of aspirin alone. The Bayer Company would, of course, agree with this, but not the manufacturers of the aspirin products in the table above.

In the case of persons who are known to be allergic to aspirin, or if unknown allergic reactions are possible as with small infants, doctors often will prescribe an aspirin substitute such as acetaminophen (Tylenol®). Evidence does exist that two aspirin tablets commonly cause the bleeding of $1/2$ to 2 mL of blood into the stomach, although healing usually takes place with apparently no after effects. Aspirin also has a special anti-inflammatory action that gives relief from the pain and swelling of arthritis, which still makes it the nonprescription drug of choice for this purpose.

In addition, aspirin has recently been shown to offer protection against some heart attacks and to reduce high blood pressure during pregnancy, although many doctors still prefer to prescribe acetaminophen instead of aspirin to pregnant women. An extensive study on over 660,000 people conducted by the American Cancer Society indicated that men and women who took aspirin at least 16 times a month had a 40% lower risk of dying from colon cancer than did nonusers.

Oil of Wintergreen

Wintergreen is a small creeping evergreen shrub common to eastern North America. A fragrant oil can be separated from the leaves of this plant and, in 1843, the chief component of this oil of wintergreen was shown to be a compound chemically classified as an ester called methyl salicylate. It was found in 1997 that certain plants like tobacco produce oil of wintergreen as a defense mechanism when under attack by viruses. Thus the role of oil of wintergreen as a plant pheromone may be just emerging.

Its similarity to salicylic acid is demonstrated by the ease with which salicylic acid can be converted into oil of wintergreen. Your experiment will carry out this transformation as illustrated in the following equation:

salicylic Acid   +   methanol   →(Sulfuric Acid Catalyst)   methyl salicylate   +   water

[The shaded boxes indicate that this conversion involves the net replacement of one hydrogen atom in salicylic acid by a methyl group (CH₃) to produce oil of wintergreen.]

Oil of wintergreen is used in perfumery and as a flavoring agent. It has a mild irritating action on the skin producing a warm sensation as blood rushes to the site. This action as a counterirritant for sore muscles explains its use in rubbing liniments. Even if its smell is not familiar to you, the odor should be pleasantly refreshing and a welcome change from many other smells prevalent in a chemistry laboratory.

The pleasant smell is not due to the close chemical relationship between oil of wintergreen and salicylic acid, but more to the fact that oil of wintergreen is a type of compound classified by organic chemists as an *ester*. *Esters* have, somewhere in their structure, carbon and oxygen atoms connected in the fashion shown in the box at right. The remaining chemical bonds (hooks in the structure) can be connected to other carbon or hydrogen atoms. Many of our most pleasant and familiar "fruity" and "flower" odors are due to esters.

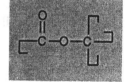

ester group

A strong catalyst like sulfuric acid (used in car batteries) is needed to force a reaction between *methanol* (methyl or wood alcohol) and *salicylic acid*. A relatively large amount of acid is used to keep the low boiling methanol from frothing out of the test tube while being heated in hot boiling water. The final addition of sodium bicarbonate neutralizes this acid catalyst with the simultaneous production of considerable bubbling due to the carbon dioxide gas formed in the reaction:

$$2\,NaHCO_3 + H_2SO_4 \longrightarrow Na_2SO_4 + 2H_2O + 2CO_2$$

| sodium bicarbonate | sulfuric acid | sodium sulfate | water | carbon dioxide |

Sodium bicarbonate (baking soda) is also the substance that can be used to neutralize excess acid on the top of car batteries in order to prevent corrosion around the battery terminals and car body.

Acidhead.

Procedure

Note #1:   You may begin the experiment with either Part A or Part B. If you are working with a partner, both parts A and B can be done simultaneously.

Note #2:   The most common problems in doing these experiments successfully are incorrect weighing of chemicals and, especially, misreading of bottle labels. **MAKE SURE YOU GRAB THE RIGHT BOTTLE!**

A. Aspirin

1. Weigh out 2.0 g of salicylic acid and place it into a dry six inch test tube. Go to a HOOD and measure out 2 mL of acetic anhydride (*CAUTION: FLAMMABLE, CORROSIVE AND POTENT ODOR. Do not get on skin*) into a dry 10 mL graduate. Pour this onto your salicylic acid sample in the test tube. Finally, toss approximately 0.4 gram of sodium acetate (CH_3COONa) into the test tube and mix well with a solid stirring rod or glass tube that has the business end sealed up.

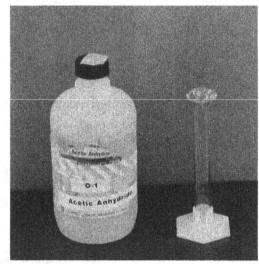

Figure 24.1. Ready to measure out acetic anhydride.

Fill a 250 mL beaker half full of water and place onto a wire gauze held on an iron ring clamped to a ring stand. Heat to boiling with a Bunsen burner. Carefully place the 6 inch test tube containing your goodies into the boiling water and simmer for 10 minutes. Mix the solution periodically during this heating period until the remaining solid has completely dissolved.

2. After heating, pour the contents of the test tube into 30 mL of water contained in a 125 mL Erlenmeyer flask. Swirl and agitate the liquid until the heavy liquid globules of unreacted acetic anhydride which sink to the bottom have completely decomposed and disappeared (under 10 minutes). Record your observations. A solid may already have started to crystallize out by this time.

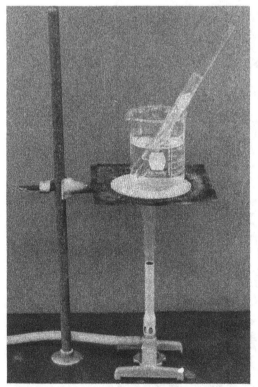

Figure 24.2. Aspirin ingredients in a test tube being heated in a water bath.

Figure 24.3. Aspirin crystallizing inside cooled flask.

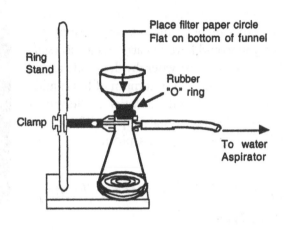

Place filter paper circle Flat on bottom of funnel

Ring Stand

Rubber "O" ring

Clamp

To water Aspirator

Figure 24.4. Suction filtration set-up.

3. Cool the flask contents by swirling in a pan or beaker of ice water. If a precipitate has not already formed, you will have to stopper the flask and shake *vigorously* as soon as the liquid in the flask starts to turn milky. If you get such a milky liquid, continue agitation and cooling until a solid precipitate has crystallized. Note your observations. If you still have difficulty getting a solid to form, leave the flask in an ice bath and proceed to Part B. Solid aspirin should have formed by the time you are finished with Part B.

Separate the precipitate by suction filtration using a small (4 cm) Büchner funnel and wash the precipitate twice with 15 mL portions of cold water.

 (This technique was discussed in Experiment 3: *Recycling Aluminum Chemically*; your lab instructor can also help you with this step. See Figures 24.4 and 24.5.) Don't forget to turn off the suction first before adding each water wash portion. After sucking the precipitate damp dry and when no more liquid drops fall from the Büchner stem, remove and spread the product out to air dry on a smooth piece of paper under a heat lamp, or place in an oven for about 10 minutes at 75 °C.

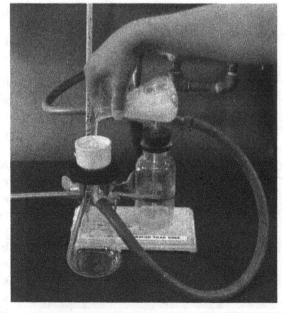

Figure 24.5. Suction filtering aspirin crystals.

4. When your aspirin is dry, weigh and record the yield (expect 1–2 g). Although fairly pure, your aspirin product probably contains a little unreacted salicylic acid. If this experiment required a very pure product, you would have to carry out a *recrystallization*—a process whereby the crude aspirin is dissolved in a minimum of hot liquid (*solvent*) and then the *solution* is allowed to cool slowly until crystals appear. These crystals would be very pure aspirin.

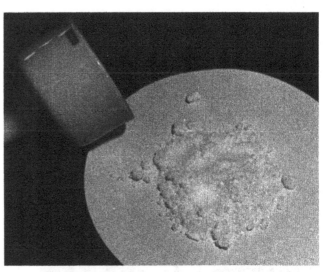

Figure 24.6. Aspirin crystals: admire!

B. Oil of Wintergreen (Methyl Salicylate)

1. Place about one-fourth of a gram (0.25 g) of *salicylic acid* into a three or four inch test tube. Then add to this test tube 40 drops (1 mL) of *methanol* (methyl alcohol—**Caution:** *flammable*). Use a stubby medicine dropper to make this dropwise addition and stir until the solid has dissolved.

You next need to add some concentrated sulfuric acid to your test tube. (**Caution:** *Very corrosive.*) If there is a "community" beaker of sulfuric acid in the hood with droppers already in it, use that. If not, **carefully** pour a little concentrated sulfuric acid into your 50 mL beaker. (Estimate about one-sixteenth of an inch of liquid in the bottom of the beaker.) If the sulfuric acid bottle is setting on a paper towel, look for its corrosive power evidenced by charred holes eaten in the paper by acid dribbles. **Rinse your hands** after using the acid bottle!

Using a medicine dropper, **carefully** add 20 drops of sulfuric acid to your sample in the small test tube with periodic stirring using a small solid glass rod. Note your observations. (Your mixture will likely turn solid but do not be alarmed if it does not.)

NOTE: If you poured a little acid into your own beaker, do not pour water onto any sulfuric acid left. Your lab instructor will advise you whether to dispose of any left over acid in the beaker by slowly pouring it into an appropriate waste container or by slowly pouring it into the sink while the water is running. Then wash out the beaker and medicine dropper with water to remove any remaining acid.

Figure 24.7. Adding methanol to salicylic acid.

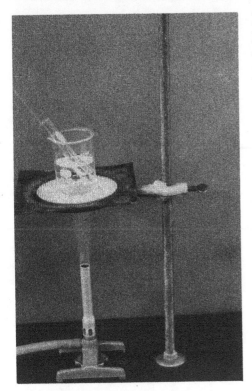

Heat the test tube and contents in a 100 or 150 mL beaker half filled with boiling water (similar to step A-1). Stir until the tube contents have liquefied and then continue heating for ten minutes. While waiting, weigh out roughly 4 g of sodium bicarbonate (baking soda, $NaHCO_3$) into a 250 mL beaker, add 50 mL water and swirl until most of the solid has dissolved.

2. After the 10 minutes total heating time has passed, pour your tube contents quickly (without cooling) into the sodium bicarbonate solution. When the fizzing subsides, stir thoroughly and cautiously smell the beaker contents. Record your *visual* and *olfactory* observations from the oil of wintergreen on the report sheet.

Figure 24.8. Heating oil of wintergreen ingredients using a boiling water bath.

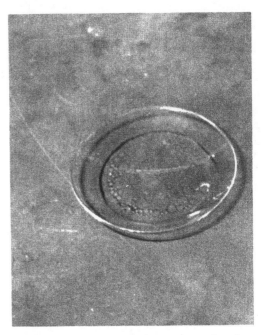

Figure 24.9. Oil of wintergreen droplets. floating on water.

Salicylic Acid Derivatives

Date _____ **Section Number** _____ **Name**_____

A. Aspirin

1. Appearance of mixture of salicylic acid, sodium acetate and acetic anhydride before heating.

2. Observations after pouring contents of heated tube into water.

3. Results from cooling the water solution of aspirin in the ice water bath.

4. Aspirin yield

    (a) Weight of crude dry aspirin .... _____ g.

    (b) Percent of maximum possible yield of 2.6 g
    (divide line 4(a) by 2.6 and multiply by 100) .... _____ %.

    (c) Appearance of crude aspirin product

B. Oil of Wintergreen

1. Appearance of mixture of salicylic acid, methanol, and sulfuric acid before heating.

2. Observations after pouring contents of heated tube into sodium bicarbonate solution.

(a) Fizzing

_____.

(b) Smell

_____.

(c) Oily droplets

_____.

Conclusions and comments on this experiment

Salicylic Acid Derivatives

Date _____ **Section Number** _____ **Name**_____

1. Draw the structure of methyl salicylate and circle just that part of the structure which classifies it as an *ester* (Hint: See the background section.)

2. How do you know that the oily globules resulting from your oil of wintergreen preparation are not simply unreacted methanol? (HINT: For help you can look up the solubility (*miscibility*) of methanol (methyl alcohol) in water in *The Merck Index*.)

3. The following six compounds are, or were, found in common nonprescription analgesic and antipyretic drugs: acetylsalicylic acid (aspirin), caffeine, salicylamide, acetophenetidine (phenacetin), salicylic acid, and acetaminophen. Refer to *The Merck Index* and look up for each the MLD (minimum lethal dose) or LD$_{50}$ (lethal dose for 50% of individuals tested). (Do not use the 9th Edition of *The Merck Index*; this edition has an error that will give you an incorrect answer.)

  (a) Which one of these drugs is the least toxic (to mice, rats, guinea pigs or rabbits)?

  (b) Which one of these drugs is the *most* toxic?

4. Aspirin tablets also contain binders which serve to hold the tablet together and reduce powdering. Assume that starch is being used for this purpose. Suggest how one might separate the aspirin from the starch. (HINT: use *The Merck Index* to look up in what liquids aspirin and starch will and won't dissolve.)

Think, Speculate, Reflect and Ponder

5. Baking powder contains sodium bicarbonate plus an acidic salt like sodium acid tartrate or calcium acid phosphate. Suggest how baking powder can "raise" breads and pastries when mixed with water and heated. (Hint: What chemical reaction would probably occur as suggested from reading the background discussion to this experiment.)

6. How do many aspirin manufacturers, whose formula for their aspirin product is exactly the same as everyone else's, make you want to buy their product instead of a competitor's product?

7. Again considering the background discussion, what statement is the artist trying to make in the sketch appearing at the beginning of this experiment?

Experiment 25

Pyrolysis of Wood

Sample from Home

A wood dowel will be furnished for this experiment. If desired, you may instead bring a similar piece of some special wood from home having *maximum* dimensions of $^7/_8$ inches (round) x 6 inches long and weighing between $^1/_2$ and 1 ounce.

Objectives

We will examine some chemistry of, and chemicals from, the renewable resource wood using the method of destructive distillation.

Background

Trees certainly rank high in importance as both a scenic and economic resource; their vital role in the pioneering development of the United States is legend. And who has not heard the words of Joyce Kilmer immortalized into the song "I think that I shall never see a poem lovely as a tree?" It thus seems fitting that we examine a little of the chemistry behind this poetic yet valuable resource.

We may not associate that common "buzzphrase" *solar power* with wood, but this renewable resource does indeed represent stored solar energy – just like the fossil fuels gas, coal and oil. Actually, the only kinds of energy that do not come either in whole or in part from the sun are geothermal, nuclear and sulfur (sulfides) and apparently hydrogen used by certain deep ocean bacteria.

Most consumers probably only think of heat and energy when asked what fuels are "good for." There exists, however, an extensive and thriving chemical industry which develops and manufactures an amazing spectrum of everyday products from these fuels — mainly coal and oil, but also wood. And wood is renewable, whereas once we have depleted our nonrenewable coal and oil reserves, so too will we have depleted our cheapest source of carbon atoms to make everything from plastics to synthetic fabrics!

Wood is a very complex substance containing roughly 50% cellulose (the *bulk* in our diet), 15% hemicelluloses, 30% lignin (acts as a matrix or binder for the cellulose fibers) and 5% "other" (mineral salts, sugars, fats, resin & protein). The actual proportions of these ingredients vary with the kind of wood, method of analysis and even the definition of the terms themselves. But in spite of this complexity, resourceful chemists have found ways to obtain directly from wood, or make from wood, a wide variety of products.

Some non-pyrolysis wood products include rayon fabrics, cellophane, cellulose based water thickeners, natural rubber, pine oil, turpentine, vanilla, camphor, maple syrup and gum arabic glues. And wood pyrolysis gives still more products. Wood is thus not just for building and burning. We will focus our *energies* in this experiment upon an historically old, but useful, process called destructive distillation *(pyrolysis)* of wood—a very special kind of "burning" as we shall see.

When we think of wood burning, our thoughts perhaps return to those memorable times we have surely all experienced when our eyes have stared long and transfixed at the uncontrolled *chemical oxidation* of logs in a fireplace. The thermal ripping apart of countless chemical bonds in the wood unleashes countless molecular fragments careening wildly towards *chemical death*. The flickering magical light display from trillions of screaming electrons stretches its luminous fingers to caress both the firebox and, seemingly, our very being. Sparks and glowing embers, throbbing with infrared radiation, ascend up the chimney—as so our thoughts are likewise carried ever higher heavenward toward . . . . . . i n f i n i t y. In our prolonged mesmerized state, we see the log reduced to a small, sad pile of dull and quite lifeless powder called wood ash.

The cycle is complete. The tree that was life to the log has now been completely consumed, and the atoms of which it was made chemically transformed into the very compounds from which it was born—carbon dioxide (gas), water (gas) and traces of minerals (ash). Ashes to ashes, and dust to dust! And the energy of the sun, chemically soaked up by the tree using photosynthesis while growing, has now been returned as heat and light. The chemical balance sheets are complete: atoms *in* (during life) equal atoms *out* (after death); energy *in* (during life) equals energy *out* (after death).

Yields of Gases from Australian Eucalyptus Wood (approximations)		
Wood Gas	@ 500 °C	@ 1000 °C
Hydrogen	1 ½ %	39%
Methane	8	9
Ethane	2½	1
Ethylene	1	1
Carbon monoxide	48	30
Carbon dioxide*	39	20
* not combustible		

Table 25.1. Wood gas analysis.

Ah, but wood does not have to burn in such a fashion. This experiment takes to an extreme what begins happening inside a wood burning stove — a *destructive distillation*. The airtight doors on such stoves severely restrict the amount of oxygen that can reach the wood to burn it. If one closes all the air vents in the stove, the wood will "simmer and cook" overnight (not a good

Pyrolysis temperature (celsius)	Percent of charcoal that is pure carbon	Percent of dry wood converted into charcoal
200	52	92
400	78	41
600	98	29

Table 25.2. Charcoal data from the destructive distillation of wood at various temperatures.

or safe practice). Although the wood will quickly and conveniently burst into flame when the air vents are opened in the morning, during the night much of the wood will have literally boiled up into the chimney or stove pipe. Not only does this partially decomposed, distilled wood deposit (called *creosote*) represent lost fuel for your fireplace, if it ever should ignite you will have a chimney fire that can sound like a jet plane is taking off right in your living room! (If you are fortunate, your house will not catch fire also.) In addition to the creosote, many unburned fuel gases during the night have also gone up and out your chimney robbing you of still more fuel that could have served to heat your house instead of polluting the air. The combustible wood gas produced contains the very poisonous carbon monoxide (CO), small amounts of methane (CH_4) and similar molecules together with a little hydrogen (H_2). You may recall how coal gasification involves converting it into methane using hydrogen gas. All wood stoves sold in the U.S. must meet federal emission standards; state laws may be even more strict. Most manufacturers have installed a built-in catalytic converter in the stove (not unlike, in principle, those in cars) which enables these wood gases to burn away, thus minimizing creosote formation and air pollution.

We have just used an example of simmering wood to describe a partial destructive distillation occurring in an (almost) airtight wood stove. If wood is intentionally heated in the total absence of air under controlled conditions, we can perform a complete destructive distillation. Heat literally causes the wood molecules to decompose into a variety of hundreds of different molecules made up only of the original atoms in the wood itself. Without oxygen from the air, wood cannot oxidize (burn) completely to carbon dioxide and water. Exactly what products are produced, and especially their amounts, are *wildly* dependent upon the type of wood used and distillation conditions. Note, for example, the effect that different temperatures have on the yield and composition of charcoal and wood gases as shown in the Tables 25.1 and 25.2.

Historically, wood was first air dried for 1–1½ years and then heated to a minimum of 270°C in kilns, at which temperature wood spontaneously starts to burn by itself. This "simmering" is allowed to raise the temperature to over 500 °C as the decomposition gases and "juices" are slowly boiled out. After a day or two, this process yields a complex mixture of products that can be processed in different ways depending upon the type of wood used and the particular chemicals being marketed. A typical overview might look like this:

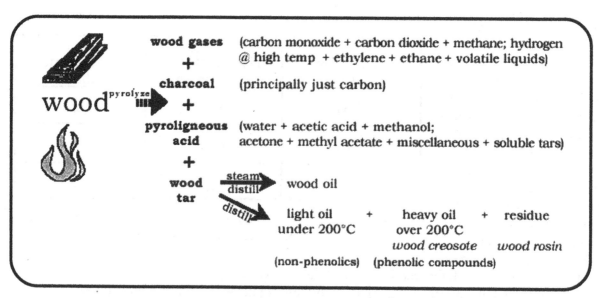

Table 25.3. Wood pyrolysis flow scheme.

The condensed liquid, called pyroligneous acid, is mostly water but commonly contains 4–10% acetic acid ("wood vinegar"), 1–6% methyl alcohol ("wood alcohol") and 0.1–0.5% acetone (nail polish remover) as the main ingredients in the molecular "soup" appearing in Table 25.4 on the next page.

Uses for some of these chemicals produced from the pyrolysis of wood have included:

pyroligneous acid .... limited use in meat smoking, leather tanning, weed killer
acetic acid .... vinegar, cellulose acetate, acetate rayon, organic synthesis
acetone .... nail polish remover, solvent, organic synthesis, acetylene absorbent
methyl alcohol .... dry cleaning liquid, fuel, organic synthesis, paints & varnishes
methyl acetone .... mixture of acetone, methyl alcohol & methyl acetate sold as a solvent
charcoal .... barbecues, industrial chemical reductions, decolorizing, tires, inks
wood tar .... rubber, paper, soaps, insecticides, fish nets
wood oil .... paints & varnishes
wood creosote .... a phenolic liquid used as a preservative, antiseptic, expectorant
wood rosin .... soaps, paints & varnishes, paper coating

acetic acid   acetone   methyl alcohol   methyl acetate

Up until about 1920, the destructive distillation of wood was an important source for acetic acid, acetone, methyl alcohol and "methyl acetone." However imports and other cheaper methods of production have greatly diminished the importance of most of these wood chemicals since that time.

This *pyrotechnic* experiment has you fire a wood sample very hot in order to complete the decomposition quickly. In less than $1/2$ hour of heating, you should have the wood changed into charcoal, wood tar, pyroligneous acid and wood gases. You will see the fuel value of the wood gases and charcoal by burning them afterwards to carbon dioxide and water (our chemical energy death). The multitudinous molecular "debris" present in the pyroligneous acid and wood tar layers will only permit our measuring their total volume and acidity. But so you may be (perhaps) impressed by the true complexity of what you have made, some of the chemicals that have been identified as mainly present in just the pyroligneous acid layer are listed in the following table:

Some Chemicals Present Mainly in the Pyroligneous Acid Layer

formic acid	pyromucic acid	methyl ethyl ketone
acetic acid	methyl alcohol	ethyl propyl ketone
propionic acid	allyl alcohol	dimethyl acetal
butyric acid	acetaldehyde	methylol
valeric acid	furfural	valerolactone
caproic acid	methyl furfural	methyl acetate
crotonic acid	acetone	pyrocatechin
angelic acid	pyroxanthene	ammonia
methyl amine	methyl formate	isobutyl alcohol
isoamyl alcohol	methyl propyl ketone	ketopentamethylene
α-methyl-β-ketopentamethylene	pyridine	methyl pyridine

Table 25.4 Pyroligneous acid molecular debris.

In addition to the pyroligneous layer, a typical commercial destructive distillation will also result in 5–13% of the dry weight of wood ending up as tars. We will spare you the list of several hundred compounds that have been identified as being present in this tar layer! Additional methods used by chemists to separate out useful products from the pyrolysis of wood include fractional distillation, extraction, steam distillation, vacuum distillation, and reaction with lime.

The residue left from burning the charcoal represents just the inorganic minerals present — largely potassium carbonate, sodium carbonate, and sodium oxide. These compounds give a strongly alkaline (basic) solution and were what *Grandma* used to break down cooking fats into her famous "lye soap" many years ago.

Extensive research on the thermal decomposition of wood has shown that some major components of wood will typically yield distinctly different types of chemical products:

beta-glucosan from cellulose   phenolic compounds from lignin

common fragment   pyrogallol 1,3 - dimethyl ether   guiacol

However the sheer number and complexity of the many compounds produced make the isolation of most of these chemicals exceedingly difficult and economically unattractive. Even today the chemical structures of the hemicellulose and lignin parts of wood itself still have not been completely established.

Procedure

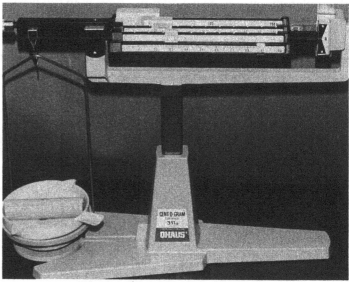

1. Place a $^7/_8$ x 3 inch wood dowel on a triple beam balance and record the weight to the nearest 0.1 g. If you have brought a wood sample from home, it should weigh between $^1/_2$ and 1 ounce (15–30 grams). But your piece of wood MUST NOT have a diameter greater than $^7/_8$ inch nor a length longer than 6 inches (cut some off with a hack saw blade if necessary).

Figure 25.1. Weighing a wood dowel on a triple beam balance.

Place the weighed wood object into the bottom of a large (8 inch) test tube. Then, using your manipulative skills and the bent glass/stopper assembly provided, assemble the apparatus as shown below and pictured "live" on the next page.

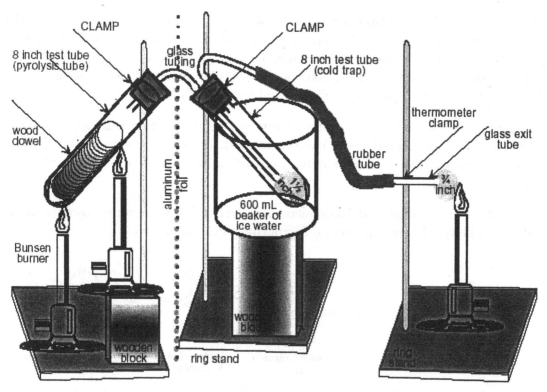

Figure 25.2. Set-up for the destructive distillation (pyrolysis) of wood.

Figure 25.3. Set-up for the Destructive Distillation (Pyrolysis) of Wood.

Wrap the pyrolysis tube clamp arm with aluminum foil before clamping it onto the tube. (This helps prevent burning the rubber coating on the clamp arm.) Don't forget to position the sheet of aluminum foil as indicated. Heed all distances noted in the procedure step 1 and have the cold trap test tube go clear to the bottom of the beaker containing ice water. There should be about $1^1/_2$ inches below the end of the delivery tube and the bottom of the cold trap test tube. You will need a total of three (3) Bunsen burners. When the engineering is complete, have it checked for "blast off readiness" by your lab instructor before proceeding. Once begun, do not stop heating until the experiment is completed. (Your delivery tube could plug up with wood tar if allowed to cool.)

2. After you have gotten the green light to go ahead, light the exit gas burner, adjust to give a small gentle flame, and position this flame about $^3/_4$ inch from the glass end of the exit gas tube. Light a second burner and adjust it to give a strong hot flame. (We'll not light the third burner until step 3.) Ask the lab instructor if you are unsure how to make these adjustments. Position the burner underneath the upper half of your wood dowel so the hot part of the flame impinges directly on the middle underside of the tube. If necessary, place a book or wood block under the burner to give it the proper height. Blast away for five minutes, recording observations on the report sheet.

NOTE: If you see the glass wall bulge where the flame hits it, cut back a little on the heat to prevent blowing a hole in the tube – a spectacular display, but not helpful towards completing the experiment!

3. After this initial five minute heating period, light a third burner, adjust to give a strong hot flame and position it under the **lower half** of your wood dowel so the hot part of the flame impinges on the bottom of the tube. (You now have two burners heating the test tube containing the wood dowel.) Continue blasting away for **another five minutes.** Periodically move the third burner away from the end of the exit gas tube to see if the gases will continue to burn by themselves. Promptly replace the burner by the exit gas tube when the wood gas flame extinguishes itself. (The gas stench and your neighbor's ire will remind you if you forget to relight the exit gases with your burner!) Record your observations.

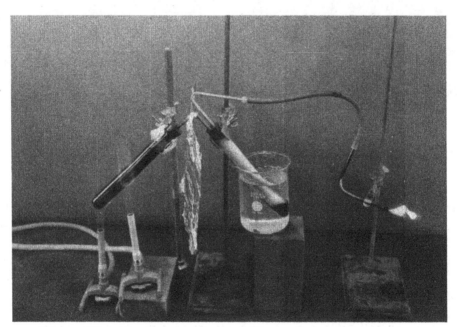

Figure 25.4. Pyrolysis underway with wood gases burning by themselves at exit tube.

4. You have at this point been heating the wood for a total of **ten minutes.** Continue heating for **another ten minutes,** but now move the burners around underneath and above the hot test tube. Hold one of the burners in your hand and direct the flame down onto and along the upper side of the test tube so as to *carbonize* any remaining wood at the top.

By the time you finish this last heating step, the wood sample will have been destructively distilled (pyrolyzed) for a total of 20 minutes and should appear completely black. The exit gases should also have ceased, as evidenced by no flame or smoke being emitted. If this is what you observe, turn off your two heating burners. If not, see your lab instructor who may have you heat your sample a bit longer. Record any observations/comments from this step on the report sheet.

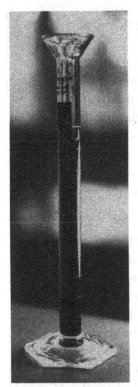

Figure 25.5. Wood distillate: 9.3 mL total volume and 1.9 mL tar layer.

5. Remove the cold trap tube containing the wood distillate liquid from your set-up. While holding with your hand, warm the contents of this cold tube by swirling over the *gentle* flame (no hot inner blue flame) from your exit gas burner. Remove the tube from the flame every few seconds and feel with your hand. Once the contents feel warm (not hot), pour all the liquid into a 10 mL graduate and temporarily set it aside to give the two liquid layers time to separate as shown in Figure 25.5. You may now turn off the last Bunsen burner.

(Proceed to step 6 at this point while waiting for your liquid layers to separate.)

Note the smell and also the color of your two liquid layers. Read the volumes of the two layers in the graduate to the nearest 0.1 mL and record all data. Also measure the acidity/basicity of the top aqueous (water base) layer by touching a strip of pH paper to it and observing the paper color change. Record pH data on the report sheet

6. While waiting for the liquid layers to separate, weigh an empty evaporating dish to the nearest 0.1 g. Then carefully — do not burn yourself — loosen the test tube clamp and remove the heated tube containing the carbonized wood (charcoal) from your set-up. By inverting and tapping the tube, "pour" the black solid contents into your weighed dish. (You can use the pointed end of a file to help dislodge any reluctant pieces of charcoal.) Reweigh the evaporating dish plus contents and record all data on the report sheet. Note the small amount (0.1–0.2 g) of solid coating left on the inside of this heated pyrolysis tube.

Now return and finish step 5 by reading the liquid volumes in your graduate.

7. Our last job will be to burn the charcoal chunks in your evaporating dish. But in order to burn up the charcoal quickly and not have to wait two hours if we just used air, we will use some pure 100% oxygen to get things *really* going. Get ready for some excitement!

The charcoal burning set-up may already be assembled for you in the hood. If not, use a ring stand and set up the simple apparatus as shown in Figure 25.7 on the next page. Place your charcoal dish directly under the inverted long-stem glass funnel and slide the funnel down until its mouth fits snugly over the evaporating dish. Carry this ring stand assembly to the hood (or wherever the oxygen supply is located).

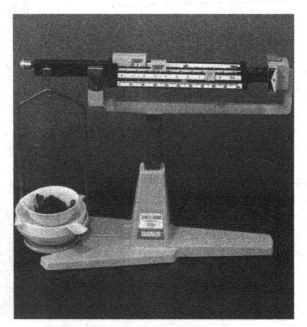

Figure 25.6. Weighing charcoal residue from pyrolysis.

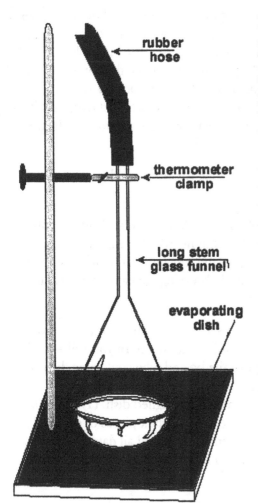

Figure 25.7. Set-up for burning
charcoal in pure oxygen

Light a Bunsen burner and adjust it to give a
strong hot flame (blue inner cone).
Temporarily slide the inverted funnel up and
away from over the dish. Hold the burner in
your hand and direct the hot flame down onto
the charcoal surface. Continue heating like
this for about a minute until the charcoal chunks have
started (glowing) — like an outdoor barbecue! Turn off
the Bunsen burner. You may *toast* your hands over
your miniature "barbie" to verify that it is indeed
working.

If not already connected, attach a 2 psi oxygen pressure hose to the top of your inverted funnel stem.
Then slide the funnel back down until its mouth again
fits snugly on top of the evaporating dish. Slowly open
the screw clamp attached to the rubber oxygen delivery tube until you see the charcoal begin to glow
brightly. Your lab instructor can assist you with this adjustment. **RESIST THE TEMPTATION** to
add oxygen so fast that the charcoal takes off
like a hundred fourth of July sparklers!

The heat from such a rapid burning will likely
shatter even a porcelain dish. Your lab instructor can help you decide what a safe, but
attractive, glow looks like.

Let the charcoal continue burning until the
glowing stops (under ten minutes) and then
shut off the oxygen flow by screwing back
down the clamp. Use tongs to carry your
evaporating dish back to your desk station.

(a) Record observations from burning.

(b) When the dish has cooled, observe
and record the nature of the residue.

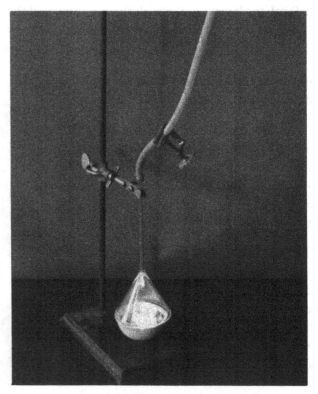

Figure 25.8. Charcoal reacting
with pure oxygen.

Experiment

Figure 25.9. Ash residue from charcoal.

(c) & (d) Reweigh the evaporating dish and calculate the net weight of ash produced from your wood sample.

(e) Add 5 mL of water to the dish contents and stir for a minute or two. Measure the acidity/basicity of the ash/water solution by touching a strip of pH paper to the liquid and observing the paper color change. Enter your results on the report sheet.

8 & 9. Use your collected data to perform some calculations on the composition of the wood sample as called for on the report sheet.

You may wish to use some acetone (Caution: *flammable liquid!*) from a squirt bottle for cleaning wood tar from your fingers, the 10 mL graduate and the 8 inch test tube that was used for the cold trap.

Return the stopper/bent tube assembly and the now blackened 8 inch test tube in which your wood was "burned" (pyrolyzed) to the location designated by your lab instructor. You may also be told to obtain a clean replacement 8 inch test tube at the same time.

Report Sheet

Experiment 25

Pyrolysis of Wood

Date _____ **Section** _____   **Name** _____

1. Weight of wood dowel .... _____ g.

2. Observations during heating top half of wood dowel

    (a) hot tube   _____

    (b) cold tube   _____

    (c) exit gases   _____

3. Observations during heating whole piece of wood

    (a) hot tube   _____

    (b) cold tube   _____

    (c) exit gases (will they burn by themselves?) _____

4. Observations/comments during final 10 minute heating _____

5. Wood distillate

    (a) smell   _____

    (b) color   _____   _____

                (top water layer)   (bottom water layer)

    (c) total volume in graduate .... _____ mL.

    (d) volume of just lower (wood tar) layer .... _____ mL.

    (e) weight of just lower (wood tar) layer .... _____ g.
            (assume density = 1.00 g/mL)

    (f) volume of just aqueous (water) layer .... _____ mL.
            (line 5c – line 5d)

    (g) weight of just water layer .... _____ g.
            (assume density = 1.00 g/mL)

    (h) pH of water layer:

     _____   _____   _____

          (color of test strip)   (pH numerical value)   (meaning of pH
value)

6. Evaporating dish

    (a) weight of dish + charcoal .... _____ g.

    (b) weight of empty dish .... _____ g.

    (c) net weight of charcoal .... _____ g.
            (line 6a – line 6b)

    (d) appearance of pyrolysis tube deposit _____

7. Burning of charcoal

    (a) effect of pure O_2 on glowing charcoal _____

    (b) appearance of ash residue:

        color _____   texture_____

    (c) weight of evaporating dish + ash   _____g.

    (d) net weight of ash ...._____g.
        (line 7c – line 6b)
    (e) pH of ash + water:

| _____ | _____ | _____ |
| (color of test strip) | (pH numerical value) | (meaning of pH value) |

8. Summary of weights obtained

    (a) charcoal ...._____g.
        (line 6c)
    (b) wood tar distillate ...._____g.
        (line 5e)
    (c) aqueous (water) distillate ...._____g.
        (line 5g)
    (d) ash _____g.
        (line 7d)
    (e) total solids/liquids ...._____g.
        (add lines 8abcd)
    (f) total gases evolved ...._____g.
        (line 1 – line 8e)

9. Percentages of original wood sample

    (a) % of wood giving charcoal ...._____%.
        (line 8a + line 1) x 100
    (b) % of wood giving wood tar distillate ...._____%.
        (line 8b + line 1) x 100
    (c) % of wood giving aqueous (water) distillate ...._____%.
        (line 8c + line 1) x 100
    (d) % of wood giving ash ...._____%.
        (line 8d + line 1) x 100
    (e) % of wood giving solids & liquids.._____%.
        (line 8e + line 1) x 100
    (f) % of wood giving gases.._____%.
        (line 8f + line 1) x 100

10. Conclusions/comments on experiment.

Pyrolysis of Wood

Date _____   Section Number _____   Name_____

1. List by chemical formula the three most abundant combustible fuel gases produced from the destructive distillation of wood and write the chemical reaction occurring when they are burned in air.

(a) _____

(b)_____

(c)_____

2. Why didn't the wood simply completely burn up in the test tube upon heating?

3. What is creosote and what problems does it cause in wood burning stoves?

4. Give a definition of the following terms:

(a) Pyrolysis

(b) Distillation

5. About half of wood is pure cellulose. What is the molecular formula for cellulose?

6. Pyroligneous acid (your top liquid layer in this experiment) contains 85-95% water. The bulk of this water was not present in the original dried wood. Where then did the atoms making up this water come from?

7. What happened to the inorganic mineral salts present in the wood, i.e., where did they finally end up after the experiment was completed?

Think, Speculate, Reflect and Ponder

8. Based on the results from this experiment, what might be a likely difference between *wood* vinegar and *wine* vinegar?

9. State two reasons why the experimental directions had you pass the wood gases through a flame.

   (a) _____

   (b)_____

10. Give a likely reason for the derivation of each of the three parts of the phrase "pyroligneous acid."

   (a) pyro   _____

   (b) ligneous   _____

   (c) acid   _____

11. What is the name of the small amount of material coating the inside of your heated pyrolysis test tube?

12. Remember our old fashioned open fireplace? . . . .
   (a) Why does it burn wood without producing significant amounts of wood tar, creosote, pyroligneous acid or combustible wood gas pollution?

   (b) What is the big dis advantage of this kind of "home heater" ? (Hint: answer is same as the big advantage of modern wood stoves.)

13. In the nineteenth century, chimney sweeps were susceptible to a rare form of cancer. Which compounds produced from the pyrolysis of wood might cause this cancer?

Experiment 26

Chemical Model Building

Sample From Home

No samples from home are needed for this experiment.

Objectives

The physical distances, angles, and number of atoms in molecules gives them their structures and helps to determine their physical state. Using commercial "stick and ball" modelling kits you will build a few simple three dimensional examples of common molecules.

Background

The sizes of atoms are so small that they cannot possibly be seen with the naked eye. In fact, even after you build one molecule with hundreds of thousands of atoms, the result is still too small to see. You need millions and millions of most molecules before you would even begin to see anything at all. However just because these objects and the arrangements of the parts of these objects are too small to see doesn't mean that they are not important.

In fact, the physical arrangements of atoms are one of the most important things that scientists have learned about the molecules that make up the matter in the universe. For instance, the arrangement of the three atoms in a water molecule give it special characteristics that make this molecule useful in dissolving salts that your body needs for life. At the same time, this same molecular structure is the reason why an oil spill takes a long time to disappear from the surface of the ocean—and why a spill like this is so damaging to the environment in the long run. The structure of the molecules that make crude oil is distinctly different from water's. Experiment 10: *Why is*

Water Harder Than Iron? describes in a little detail what happens when a molecule has both of these structural attributes, those of water on one end and those of oil on the other; it forms a soap or detergent.

Finally, the arrangement of the atoms in the molecules in the genetic material in my DNA also determines whether or not I have red hair. In short, the major differences between two different molecules are, of course, the kind of atoms that are present, but also the arrangement of the atoms that make up those molecules.

The simplest molecules are made of at least two atoms joined together in such a way that the atoms share two electrons. We call this process of sharing two electrons a bond. A very simple example of this is the two-atom molecule hydrogen, H_2. This molecule is referred to as *di*atomic because it has two atoms. It is also designated a *homo*nuclear diatomic because both of its atoms are the same and, of course, the atoms are identical in size and mass. The first molecule that you will build in this lab is diatomic hydrogen. This molecule is very light and has a very even (symmetrical) distribution of the electrons around the atoms that make it up. This molecule is referred to as a nonpolar molecule because of the kind of atoms, identical on each end of the bond, and because of the distribution of electrons or electron density among the atoms.

If one of the atoms of molecular hydrogen were replaced by nature (or by an inquisitive student) with a chlorine atom, things change. The second molecular model that you will build for this laboratory experiment will be HCl. This diatomic molecule has neither identical atoms nor a symmetrical electron distribution now. The atoms still share electrons but two things (in our simple treatment) have changed: First the size of the atoms have changed (a chlorine atom is much larger than an atom of hydrogen), and secondly, the distribution of electrons has changed too; many more electrons are concentrated near the chlorine atom than near the hydrogen atom. The number of electrons that atoms bring to molecules is just equal to their atomic number appearing in the periodic table (1 for H and 17 for Cl); however we will not spend too much time concentrating on the electrons for this experiment. Chemists label HCl a *hetero*nuclear diatomic and also call this type of molecule polar because of, in part, the unequal electron distribution.

The next molecular model that you will build contains just two different kinds of atoms (similar to HCl), but instead of just two atoms in the molecule, there are five atoms. This molecule is methane, CH_4. The key to understanding the structure of this molecule is to stop thinking "on the page" and to start to think in three dimensions. After all, the universe is not a two dimensional thing and neither are molecules.

Methane, CH_4, can of course, be drawn on the page like H_2 and HCl. But this is the first time that the drawings on the page really don't adequately represent reality. Instead of methane looking like that two dimensional cross in a square—with a hydrogen on each side of the square and (a larger) carbon atom in the middle—the truth is that the physical forces that make up the CH_4 molecule make the structure a *three dimensional shape* called a tetrahedron. The tetrahedron has four sides. Each side has a face like a triangle. If you think of the methane

molecule as being inside of the tetrahedral shape, then the carbon atom would be in the center and the four hydrogen atoms would each be at a vertex or corner of the tetrahedron (there are four vertices too).

The reason why the three dimensional methane molecule is not flat, with all the atoms in the same plane as in BF_3, is because of the attractions and repulsions between the forces in the molecule. More specifically, the electron pairs in the bonds that make up the molecule stay as far away from each other as possible (see Question 7 under Think, Speculate, Reflect and Ponder). In methane, these repulsions are most successfully balanced, with the lowest overall "conflict" being represented by a tetrahedral structure, not a planar "cross." In BF_3 a trigonal planar structure satisfies these requirements best.

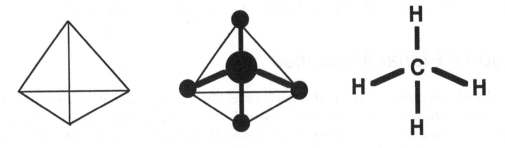

If you are starting to think that the amount of effort necessary to describe with words and draw these 3D molecules is getting out of hand, you are right. It is much easier to build them using chemical model kits and that is exactly what we will do in this experiment.

Your lab instructor has kits for you to use and will tell you how to put the models of the molecules together. Please be careful with the kit's atoms and the bonds since they are generally somewhat fragile and have to be connected and disconnected over and over again. Some models have different colors to represent different kinds of atoms, and some have different length bonds; usually any bond to hydrogen is shorter than bonds between other atoms.

Procedure

A. Diatomic Molecules

The first molecule you will build is hydrogen. Connect two identical atoms together using the shortest bond available. There may not be any different bond lengths available; however, most molecular modelling kits do have different size or color atoms. Use the kit's instructions or check with your lab instructor—hydrogen is usually the most numerous atom in the kit. *Hydrogen can only form one bond*; this is an important rule. Other atoms like carbon can form more than one bond. In this lab we will assume that *carbon always forms four bonds*.

Notice that the H_2 molecule that you built is symmetrical. If you passed a mirror (a pane of glass) perpendicularly through the bond between the atoms with each hydrogen atom completely on either side of the mirror, then the reflection of an atom (and half of the bond left on that side of the mirror) would look exactly like the whole molecule without the mirror present.

Passing the mirror test for symmetry means that one side of the molecule is an exact mirror image of the other half. Since there are only two atoms in this molecule it is also automatically linear—it has no chance for another shape.

Hydrogen chloride is built exactly like H_2 except that one of the atoms is a different color and/or size than hydrogen. Use the atom specified by the modelling kit or your lab instructor for chlorine (or any halogen); don't use the kit's carbon atom. Remove one of your H atoms from H_2 and build HCl. (Like H, *chlorine only forms one bond*.) The fat side and thin side of the HCl molecule (that is, the large and small atoms on either side of the bond) show that the mass and electrons in the molecule are not symmetrically distributed and therefore there is more electron density at one end of the bond than at the other. This *polar* molecule would not pass the imaginary mirror image test if the mirror were passed directly between the two atoms because the molecule is not symmetrical. The reflection in the mirror (on either side) does not look like the original molecule.

2. Trigonal Planar Molecules

Though relatively uncommon compared to linear or tetrahedral molecules, there *are* molecules that have a central atom with three connected atoms all lying in the same plane, that is, flat as if they were all lying on a page. One of these molecules is BF_3. Build boron trifluoride with three fluorine atoms connected to the central boron atom. Note that you cannot use the tetrahedral atoms (see below). For the three fluorine atoms use an atom like chlorine that you used in HCl, something that is larger than hydrogen and can connect to only one other atom. The strictly tetrahedral connectors will *not* allow you to make a molecule with three atoms connected to the central atom all in the same plane. (Hint: The angle between each of the three bonds in this molecule is 120 degrees.)

3. Tetrahedral Molecules

Now you will build your first tetrahedral molecule. Methane is often used as the primary example of this shape of molecule, and we will also start here. Build CH_4 using one tetrahedral connector (the central carbon atom) and four hydrogens. Remember that carbon forms four bonds and hydrogen only one. Make sure that you use the carbon atom as specified by your lab instructor or 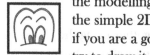 the modelling kit's instructions. Notice that the molecule is inherently NOT FLAT like the simple 2D drawing of methane in this lab's written introduction. As a matter of fact, if you are a good artist you might put the methane molecule down on the table top and try to draw it. (See the report sheet.) Estimate the angle between two bonds in methane; that is, what is the angle made between three points made up of a hydrogen atom, carbon, and another hydrogen atom? Answer the first question in the Think, Speculate , Reflect and Ponder section (Question 6) now. Notice that it is smaller than that of the trigonal angle but larger than 90º. What would you estimate that angle to be?

The reason for the bond angles we have seen in the trigonally (120º) and tetrahedrally bonded central atom has to do with electron repulsion. The electrons (all negatively charged) in the bonds are trying to stay between the (positively-charged) atoms in the bond that they are attracted to, yet stay as far away from each other as possible, because electrons in different bonds repel each other. If the central atom in methane had 4 hydrogens bonded to it and their bonds were all in the same plane the result would be four, 90º angled bonds. By existing in three dimensions, the bonds can increase their distance from other electrons in other bonds. That's the secret!

Other alternative tetrahedral molecules can be constructed by exchanging one or more of the hydrogen atoms for other atoms. Using the same kind of atom that you used for chlorine in HCl (and other than what you used for carbon or hydrogen), construct each of the following molecules in turn: CH_3Cl, CH_2Cl_2, $CHCl_3$, and CCl_4. What is the difference between CCl_4 and CH_4?

A second "level" of tetrahedral molecules can be made by joining two tetrahedral (carbon) centers together to form a longer molecule. If the only atoms used are hydrogen and carbon, these molecules are simply called hydrocarbons. Try removing one hydrogen from each of two different methane models and joining those two (CH_3-) pieces together. Most modelling kits have a longer bond connector for bonds that are not connected to hydrogen. Use the longer bond length if you can.

You have now built ethane, variously written as C_2H_6 or CH_3CH_3 or sometimes H_3C—CH_3. These different ways of writing the formulae help describe which atoms are connected together. Again notice that each carbon has four bonds and each hydrogen has only one; however, now one of the bonds is a carbon/carbon bond instead of a carbon/hydrogen bond.

Can you build the next member of this hydrocarbon family, propane, with three carbons? The carbons are "all in a row" and the center carbon is connected to one carbon on either side. All the other bonds are connected to hydrogen, and as before, each carbon has four bonds and each hydrogen only one.

Molecular Model Building

Date _____   **Section** _____   **Name**_____

1. Linear models

List the linear diatomic molecules that you built. Write down their chemical names and their formulas.

2. Trigonal planar models

What is the chemical name and formula of the trigonal planar molecule that you built? Do the best you can at drawing this molecule in the space below. Use letters to represent the atoms and lines for the bonds.

3. Tetrahedral models

List the chemical formulae of all of the tetrahedral molecules that you built that contained one carbon. Draw methane in the space below to the best of your ability. Try to make your drawing look as three dimensional as possible on the page. You will <u>not</u> be graded on your artistic ability.

4. "Joined" tetrahedral structures

List the name and formulae of two of the models that you built that had two or more carbons. Write each of the formulae two different ways. Make sure that at least one formula for each shows *which carbons are connected together.* Draw two-dimensional pictures of ethane and propane.

5. Butane

Write down the formula of the next logical member in this series of hydrocarbons, butane. This molecule has four carbons in a row and follows the same pattern that you saw for methane, ethane, and propane. How many carbons are necessary if you follow the pattern started with ethane?

6. The Rule of 90 *

Simple hydrocarbons with chlorine and fluorine replacing some or all of the hydrogens are the basis of chemicals called chlorofluorocarbons (surprise). Commercially for a long time they were called Freons® and were used in refrigerators and air conditioners as working fluids. Working fluids are liquids or gases which are used *to move heat* through a system using a series of compression and expansion steps. The simplest and most widely used chlorofluorocarbon has the formula $CFCl_3$. Notice that like methane it has four bonds to the central carbon; fluorine and chlorine never double bond. This compound is named CFC-11 using the Rule of 90. Here's how it works.

Take the CFC number and add 90 (ergo the Rule of 90): 11 + 90= 101

*For an additional description of the Rule of 90 see the entry for chlorofluorocarbons in **The Merck Index.**

That resulting sum represents the number of the atoms in the molecule: the three digit number, 101, represents the number of carbon, hydrogen, and fluorine atoms in the molecule— always in that order. Whatever bonds to the carbon are left over (remember carbons always have four bonds) are connected to chlorine atoms (remember, no double chlorine or fluorine bonds are allowed).

Therefore 101 represents a CFC with 1 carbon, 0 hydrogens, and 1 fluorine. All the rest of carbon's bonds are made to different chlorine atoms. That leaves three unfilled carbon bonds so there are three chlorines.

So the formula of CFC-11 is $CFCl_3$.

Here's one more: The formula of CFC-140 is $C_2H_3Cl_3$. Why?

Molecular Model Building

Date _____   **Section** _____   **Name**_____

1. What is the term given to two electrons shared by two atoms?

2. Give examples (name and formulae) of three different *homo*nuclear diatomic molecules using only atoms from Group VII of the periodic table. What is the angle between the bonds in all of these molecules? What was the angle between the bonds in the trigonal planar molecules that you built in Step 2 of the lab procedure?

3. Give examples (name and formulae) of three different *hetero*nuclear diatomic molecules using hydrogen and atoms from only Group VII of the periodic table.

4. An even, symmetrical distribution of the mass and the electron density in a molecule creates a _____ molecule.

5. Polar diatomic molecules are made of two _____ atoms joined by a bond.

Think, Speculate, Reflect, and Ponder

6. Build the methane molecule with your modelling kit and get an approximate angle between any two of the bonds in this molecule. In other words, what is the angle made between three points made up of a hydrogen atom, carbon atom, and a second hydrogen atom? Is the angle less than or more than 90°? What is the largest angle possible between four bonds in a three dimensional molecule like methane?

7. If the angle that you measured in Question number 6 were greater than 90°, do you think that that would decrease the repulsion between the electrons in two adjacent bonds or increase it when compared to a bond angle of 90°? Why? What is the "advantage" to the molecule of having bonds farther apart as opposed to closer together?

8. Write down the correct formula for each of these compounds using the Rule of 90:

CFC-11   CFC-12   CFC 113

CFC-123   CFC-141   CFC-10

Experiment 27

The Last Day in the Lab:
Glass Etching
Checkout

"The time has come the walrus said
To talk of many things
Of fluorine and glass and paraffin wax
Of chemistry and kings"

Sample From Home

(Optional.) Bring a small glass object on which to etch a design, name, etc.
Glass slides (2" x 3") will be furnished in lab for etching.

Objectives

The ability of hydrofluoric acid to attack glass will be demonstrated by etching a glass object which can then be taken home as a souvenir of the course. The experiment takes only ten minutes and can be done after you have checked out of your lab drawer.

Background

Of all the elements, fluorine and its compounds are perhaps the most interesting because of the unusual nature of their properties. These properties can range from the highly reactive and toxic elemental fluorine itself—the most reactive nonmetal and the first to react with the "inert" gases—to the insidiously toxic perfluoroisobutylene. On the other hand, the perfluoroethylene polymer called Teflon® is exceedingly inert and nontoxic. Fluorine plays important physiological roles that are beneficial to life (fluorides for proper bone and especially tooth growth). Also notable is the fact that some nerve poisons used as war gases and insecticides contain fluorine.

| fluorine gas | perfluoro-isobutylene | Sarin nerve gas (1 mg or 1/200 of a drop lethal to humans) | Teflon polymer (chain section) | fluoride ion |

The distinctive and unusual behavior of fluorine can be seen in the group of common hydrogen halide gases: hydrogen fluoride (HF), hydrogen chloride (HCl), hydrogen bromide (HBr) and hydrogen iodide (HI). When dissolved in water, these four gases will produce respectively hydrofluoric acid, hydrochloric acid, hydrobromic acid and hydroiodic acid. Although hydrofluoric acid (HF dissolved in water) is the *weakest* acid of this group, it is *strongest* and unique in its ability to attack readily the relatively inert substance glass. Very strong bases like lye are the only other common chemicals able to do this—and they do so only very slowly.

Because of this property, hydrofluoric acid cannot be stored in glass containers like almost all other reagents. Originally it came stored in wax containers, but now plastic bottles are cheaper and safer. Interestingly, pure gaseous hydrogen fluoride can be shipped and stored in metal cylinders if bone dry (recall how rusting requires the presence of water also).

Ordinary glass is composed of approximately three-fourths silica (silicon dioxide, SiO_2). Silica in the pure state is known as quartz or "crystal" to hikers. But the one-fourth "other" chemicals dissolved in the quartz both lower its melting point and also prevent the quartz from crystallizing—producing instead a "glassy" solid. (This impurity effect on the melting point is a general rule in chemistry.) Thus, apart from its relative chemical inertness, glass is really just a very *viscous* ("thick") liquid much like clear Karo® syrup or honey.

The physical property that makes glass somewhat special is that room temperature causes it to become so "cold" that it will not measurably flow any more. We might say that "Room temperature is to glass what a cold January morning would be to Karo® syrup—it sets up 'like glass'!" Once one realizes that glass is just an extremely viscous clear liquid, it is not so hard to understand why it is transparent and can be softened by heating and then formed into different shapes. (It is said that the bottoms of old church window panes are slightly thicker than the top. Theories about this are controversial, but can you suggest one possible reason why?)

When hydrofluoric acid and glass are brought together, the glass surface is eaten away (etched) due to chemical reactions such as:

$$4HF + SiO_2 \rightarrow SiF_4 + 2H_2O$$

$$6HF + Na_2SiO_3 \rightarrow SiF_4 + 3H_2O + 2NaF$$

$$6HF + CaSiO_3 \rightarrow SiF_4 + 3H_2O + CaF_2$$

| hydrogen fluoride | component in glass | silicon tetrafluoride | water | a fluoride salt |

The gaseous HF easily dissipates, and the sodium and calcium fluorides formed crumble and wash away leaving the pitted, etched glass surface. Frosted and etched glass can be produced commercially using HF vapor, but in this experiment we will use aqueous hydrofluoric acid (HF dissolved in water).

In order to etch only desired areas, the glass object is coated with paraffin wax which forms an inert barrier between the acid and glass surface. Etching can then only occur where this protective coating is scraped away from the glass surface.

Procedure

CAUTION: HYDROFLUORIC ACID IS VERY DANGEROUS. Your lab instructor will use thick rubber gloves and place your prepared glass object into the etching bath, take it out after the reaction has proceeded far enough (30 - 60 minutes) and then wash it thoroughly with water before you pick it up.

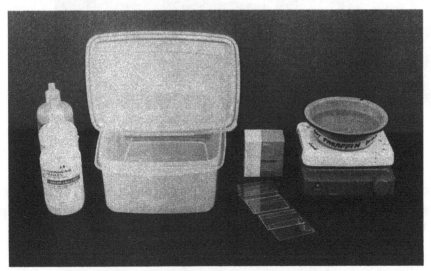

Figure 27.1. Etching bath, hydrofluoric acid, glass slides and liquid paraffin on a hot plate.

Coat a microscope slide or some other small glass object of your choice with wax by dipping it into a container of hot liquid paraffin. Holding one end with your fingers, try dipping one-half in at a time, let it cool and solidify for 10–20 seconds, and then dip in the other half. The entire glass surface must be so coated; even glass which does not actually touch the liquid must be covered since acid *fumes* can also cause etching to take place. The insides of hollow objects like cups must be similarly coated for the same reason.

Using a pen, pencil or any sharp object, write or draw anything you wish onto your coated glass by scraping away all the wax clear down to the glass surface with the sharp object. Because glass is transparent, it is generally best to write only on one side. When finished, leave it for your lab instructor to carry out the actual etching process.

When you pick up your etched glass, **please do not scrape the wax off into the sink**. One way to remove the wax from your finished etched slide or object is to hold it over a paper towel while warming gently with a *cool* type flame from your Bunsen burner (absence of hot inner cone as described in Experiment 1: *The Ubiquitous Bunsen Burner*). Most of the wax will drip off onto the towel, and you can complete the job by scrubbing off the remaining paraffin in hot water.

Figure 27.2. Some etched slides and glassware.

You now have a complementary memento of your experience in this course. The etchings may be colored by rubbing a colored pencil across the marks. As an added bonus, there are no report sheets or answers to questions to be completed for this experiment. Enjoy.

Credits

All pictures taken by the authors.

Picture assistants: Becky Richardson and Mercy Iverson.

The authors wish to thank The Weyerhaeuser Company and the staff at the Technical Center Library for making their facilities available.

Page 2 **Sketch:** Chemist caricature drawn by Lorain Stowe, Highline Community College, Seattle, Washington.

Page 19 **Stanza:** From an early anti-metric theme song "A Pint's a Pound the World Around."

Page 24–25 **CRC Tables:** Reprinted with permission from CRC Handbook of Chemistry & Physics, 51st Edition, 1970–71. Copyright © CRC Press Inc., Boca Raton, Florida.

Page 32 **Aluminum:** Sketch reprinted with permission from Contemporary Chemistry, E.A. Walters and E.M. Wewerka, Merrill-MacMillan Publishers. Copyright Edward A. Walters.

Page 34 **Ivory Soap:** Historical notes from Chemical and Engineering News, April 18, 1994, p. 64.

Page 48 **NTS radiation venting:** New York Times, Feb. 1, 1989, p. A-11. See also: Hidden Dangers: Environmental Consequences of Preparing for War, A. E. Ehrlich and J. W. Birks, Eds. Copyright © (1990) Sierra Club Books.

Page 49 **Table:** Some data from National Safety Council's Environmental Health Center (www.nsc.org/ehc.htm); Chernobyl data taken from National Geographic August 1994, p. 114.

Page 57 **Content of Air:** Based on "A General Chemistry Experiment for the Determination of the Oxygen Content of Air," by Birk, J. P.; McGrath, L.; Gunter, S. K.; Journal of Chemical Education 58(10), 804-805 (1981).

Page 70 **Natural Pigments:** Based on "An Inexpensive and Quick Method for Demonstrating Column Chromatography of Plant Pigments of Spinach Extract," by Mewaldt, W.; Rodolph, D.; Sady, M. in J. Chem. Ed. 62(60) pages 530-531(1985).

Page 83 **Combustion:** Based on "Heating Values of Fuels," by Rettich, T. R.; Battino, R.; Karl, D. J. in J. Chem. Ed. 65(6), 554-555(1988).

Page 112 **Evaporation:** "Volatile molecules" drawn by Todd Remick, Highline Community College, Seattle, Washington.

Pages 117–118 **Tables:** Reprinted with permission from CRC Handbook of Chemistry and Physics, 33rd Edition, 1951–52. Copyright © CRC Press Inc., Boca Raton, Florida.

Page 128 **Table:** Information from U. S Geological Survey and departments of public health.

Pages 140–141 **Story of Bread:** From Terracide by Ron M. Linton. Copyright © 1970. Reprinted by permission of Little, Brown and Company.

Page 148 **Chlorine:** Sketch reprinted with permission from Contemporary Chemistry, E.A. Walters and E.M. Wewerka, Merrill-MacMillan Publishers. Copyright Edward A. Walters.

Pages 154–155 **Tables:** Information taken from United States Department of Agriculture, Food Safety and Inspection Service, Meat and Poultry Inspection Operation, FSIS Directives; also Consumer Reports issues mentioned on these pages.

Page 169 **Table:** Compiled information reproduced through the courtesy of ESHA Research, Salem, Oregon (1994).

Page 170 **Indophenol Analysis:** Official Methods of Analysis, AOAC, Section 39.040.

Page 180 **Cartoon:** drawn by Todd Remick, Highline Community College, Seattle, Washington.

Page 180 **Spicy Spray:** Seattle Times, September 2, 1992 by Helen E. Jung

Page 183 **Merck Index:** Entries taken from The Merck Index, Eleventh Edition (1989), S. Budavari, M. J. O'Neil, A. Smith, P. E. Heckelman, Editors, by permission of the copyright owner, Merck & Co., Inc., Rahway, N. J., U.S.A. © Merck & Co., 1989.

Page 186 **GRAS List:** Selected and abridged from Code of Federal Regulations, Food & Drugs, Office of the Federal Register (1985).

Page 194 **Artificial Colors:** Based on "Reverse-Phase Separation of FD & C Dyes Using a Solvent Gradient," by Ondrus, M. G. in J. Chem. Ed. 62(9), 798-799(1985).

Page 202 **Lead:** Sketch reprinted with permission from Contemporary Chemistry, E.A. Walters and E.M. Wewerka, Merrill-MacMillan Publishers. Copyright Edward A. Walters.

Page 203 **Table:** Abridged from Statistical Abstracts of the U.S., p. 203 (1990); U.S. Dept. of Commerce, Bureau of the Census.

Page 204 **Cadmium:** Reprinted with permission from Chemical & Engineering News September 8, **38**, 1986. Copyright © 1986 American Chemical Society.

Page 219 **Caffeine:** Reprinted with permission from View Magazine Jan/Feb 1988, page 28. Copyright © (1988) Group Health Cooperative, Seattle, Washington.

Page 232 **Water Fluoridation:** Table 1—United States Public Health Service.

Page 233 **Food Fluorine:** Table 2 reprinted from Public Health Reports Vol. 64, page 1061 (1949) by F. J. McClure.

Page 234 **Tooth Mortality:** Table 4 reprinted from Public Health Reports Vol. 66, page 1389 (1951) by A. L. Russell and E. Elvove.

Page 234 **Tooth Mottling:** Table 3 reprinted with permission from Fluorine Chemistry, Vol. IV, p. 449 by H. C. Hodge and F. A. Smith. Copyright © (1965) Academic Press.

Page 234 **Natural Fluoride:** Table 5 from the National Institute of Dental Research.

Page 235 **Fluoridation:** Table 6 from the Centers for Disease Control.

Page 235 **Quote:** reprinted from Chemical and Engineering News March 26, 1990, page 3 by permission of Margaret M. Gemperline, Greenville, N.C.

Page 236 **World Fluoridation:** Table 7 reprinted with permission from Chemical and Engineering News Aug. 1, 1988, page 30. Copyright © (1988) American Chemical Society.

Page 237 **Thirsty Sheep:** drawn by Todd Remick, Highline Community College, Seattle, Washington.

Page 255 **Molecular Weight of Air:** "The Density and Apparent Molecular Weight of Air" by Harris, A. D. in J. Chem. Ed. 61(1), 74-75(1984).

Page 297 **Acidhead:** Sketch reprinted with permission from Contemporary Chemistry, E.A. Walters and E.M. Wewerka, Merrill-MacMillan Publishers. Copyright Edward A. Walters.

Page 308 **Wood Gases:** Data for table abridged from Hanley, N. A., and Pearse, J. F., Australian Chem. Inst. J. and Proc., 12, 263–72 (1945). See also: Wood Chemistry, Volume 2, Chapter 20, The Thermal Decomposition of Wood by A. W. Goos, p. 845, edited by L. E. Wise and E. C. Jahn. American Chemical Society Monograph Series, copyright © (1952) Reinhold Publishing Corporation.

Page 309 **Charcoal:** Data taken from Bergström, H. O. V., G. Wessland, Om Träkolning, Sweden, 1918. See also: Chemical Processing of Wood by A. J. Stamm and E. H. Harris, p. 442. Copyright © (1953) Chemical Publishing Company, Inc.

Page 311 **Chemicals:** List taken from Hardwood Distillation Industry, Circular #738, U. S. Dept. of Agriculture, Forest Service (1956). See also: Distillation of Resinous Wood, Circular #496, U. S. Dept. of Agriculture, Forest Service (1958).

Page 335 **Walrus:** paraphrased from Alice in Wonderland (with apology to Lewis O. Carroll).

Notes

Notes

Notes